EVALUATING THE UPGRADING OF TECHNICAL COURSES AT TWO-YEAR COLLEGES:

NSF'S ADVANCED TECHNOLOGICAL EDUCATION PROGRAM

ADVANCES IN PROGRAM EVALUATION

Series Editor: Robert E. Stake

Volume 6: Telling Tales: Evaluation and Narrative – edited by Tineke Abma

Volume 7: Visions of Quality: How Evaluators Define, Understand and Represent Program Quality – edited by Alexis P. Benson, D. Michelle Hinn and Claire Lloyd

Volume 8: School-Based Evaluation: An International Perspective – edited by David Nevo

ADVANCES IN PROGRAM EVALUATION VOLUME 9

EVALUATING THE UPGRADING OF TECHNICAL COURSES AT TWO-YEAR COLLEGES:

NSF'S ADVANCED TECHNOLOGICAL EDUCATION PROGRAM

EDITED BY

FRANCES LAWRENZ
Department of Educational Psychology,
University of Minnesota, USA

NANETTE KEISER
Executive Director, Berrien Community Foundation, USA

ARLEN R. GULLICKSON
Western Michigan University, USA

2004

ELSEVIER
JAI

Amsterdam – Boston – Heidelberg – London – New York – Oxford
Paris – San Diego – San Francisco – Singapore – Sydney – Tokyo

ELSEVIER B.V.
Sara Burgerhartstraat 25
P.O. Box 211
1000 AE Amsterdam
The Netherlands

ELSEVIER Inc.
525 B Street, Suite 1900
San Diego
CA 92101-4495
USA

ELSEVIER Ltd
The Boulevard, Langford
Lane, Kidlington
Oxford OX5 1GB
UK

ELSEVIER Ltd
84 Theobalds Road
London
WC1X 8RR
UK

First edition 2004

British Library Cataloguing in Publication Data
A catalogue record is available from the British Library.

ISBN: 0-7623-1120-7
ISSN: 1474-7863

The paper used in this publication meets the requirements of ANSI/NISO Z39.48-1992 (Permanence of Paper). Printed in The Netherlands.

CONTENTS

STAFFING

Principal Investigator and Project Director	Arlen Gullickson
Co-Principal Investigator and Senior Associate	Frances Lawrenz
Senior Research Associate and Project Manager	Nanette Keiser
Issue Paper and Chapter Authors	Norman Gold
	Arlen Gullickson
	Nanette Keiser
	Frances Lawrenz
	Thomas Owens
	Karen Powe
	Lester Reed
	Gloria Rogers
	Gloria Tressler
Reviewers of Issue Papers	Thomas Bailey
	John Fraire
	Charles Moore
	Albert Schwabenbauer
	Nick Smith
	Leonard Sterry
	Naida Tushnet
Editorial Assistant	Sally Veeder

We gratefully acknowledge the assistance and cooperation of the staff members of the 13 ATE projects/centers visited in 2000 and 2001, and the valuable information received from those projects/centers that responded to our surveys.

The Editors

ABSTRACT

NSF-ATE program officers were interested in widely disseminating the evaluation data collected for the ATE program. Although an overall merged evaluation report could have been produced from the survey and site visit evaluation data, Western Michigan University's (WMU) evaluation project advisory panel, in conjunction with NSF, believed that a more varied and unique approach to report writing should be tried. The intention was to more fully explore the available data through an individual, issue-oriented approach. Toward that end, the WMU evaluation project commissioned the following nine issue papers in early 2001.

OVERVIEW

OBSERVATIONS FROM THE ATE PROGRAM

The National Science Foundation's (NSF) Advanced Technological Education (ATE) program is a diverse and dynamic set of *projects*[1] intent on improving our nation's technical workforce (see pages 7–9 for detailed information about this program). To meet this intent, the ATE program is helping to provide economic empowerment for many who otherwise may have been passed by traditional programs. The following observations describe some of the impacts of this program.

- An individual completed her associate's degree in biotechnology, a first-generation college graduate, largely due to the encouragement of a mentor employed in this field.
- The ATE *project* at Pasadena City College "opened my eyes to all the different types of jobs out there for a graphic designer. It not only prepared me with great multimedia pieces for my portfolio, it also taught me how to work in a project-based environment assuming the different types of roles a project could require, as well as how to work with people in a team environment."[2]
- A decrease in truancy rates occurred for at-risk high school students in an ATE *project* because what they are studying in high school "relates to the real world – I can get a job, a good job, from what I'm learning right now" according to one of the students.[3]
- Faculty members received cutting-edge training in e-commerce through the NorthWest Center for Emerging Technologies (Bellevue Community College, Bellevue, Washington). Training was available for high school teachers and

Evaluating the Upgrading of Technical Courses at Two-Year Colleges:
NSF'S Advanced Technological Education Program
Advances in Program Evaluation, Volume 9, 1–12
Published by Elsevier Ltd.
ISSN: 1474-7863/doi:10.1016/S1474-7863(04)09001-5

community college, 4-year college, and university faculty.[4] Business/industry partners contributing to this effort included Microsoft and Boeing.

- A mother proudly pointed out that her daughter just began a technician program at the local community college because her mother finished her associate's degree in a similar field.
- Resources were leveraged through productive collaborations of business and industry, educational institutions, and other organizations. For example, the Validation and Implementation of a Coordinated Precision Agricultural Curriculum project (Hawkeye Community College, Waterloo, Iowa) not only received matching equipment funds from its participating schools, but also donations of funds, equipment, software, and other items. John Deere and Company provided $250,000 for a technologically advanced computer lab, with matching funds from the College of $350,0000. Environmental Research Systems, Inc. donated software valued at $125,000; and Rockwell International provided $200,000 in GPS equipment, software, and technical support.[5]
- A man in his late 30s shared that he is finally excited about going to work each day now that he has up-to-date job skills that made it possible for him to obtain a job he really enjoys.

These anecdotal examples provide glimpses of the range and power of the ATE program. The purpose of this chapter is to provide background information to form a platform from which to examine the ATE program in a more in-depth, issue-oriented fashion. We begin by describing the issue-papers development process. This is followed by the synopses of the various issue papers included in Chapters 2–8 and a brief description of the ATE program and its evaluation.

DEVELOPMENT OF THE ISSUE PAPERS

NSF-ATE program officers were interested in widely disseminating the evaluation data collected for the ATE program (see page 9 for detailed information about the Western Michigan University [WMU] evaluation project and methodologies). Although an overall merged evaluation report could have been produced from the survey and site visit evaluation data, the WMU evaluation project advisory panel, in conjunction with NSF, believed that a more varied and unique approach to report writing should be tried. The intention was to more fully explore the available data through an individual, issue-oriented approach. In other words, instead of a ponderous report containing all available data, smaller, more focused papers concentrating on relevant topics or issues would be produced.

Toward that end, the WMU evaluation project commissioned nine issue papers in early 2001. An issue is defined here as a topic or situation that requires further consideration to make a decision or reach a conclusion. Five issues addressed by the ATE evaluation come from the core program objectives or drivers mandated by Congress as described on pages 7–9.

- Collaboration.
- Dissemination.
- Materials development.
- Professional development.
- Program improvement.

Four additional issues evolved from discussions and concerns expressed by NSF program officers and the WMU evaluation project staff:

- Advisory committees.
- Evaluation.
- Recruitment and retention.
- Sustainability.

Authors were selected who had expertise in the various issues and were members of the existing advisory groups of the WMU evaluation project or evaluation project staff. Most authors had also participated in at least one site visit and were working with the ATE *projects* in other ways as well (e.g. evaluator for a *project*).

The authors were provided with a general outline as guidance and all the survey[6] and site visit[7] data. The authors provided additional insights through literature reviews, their personal involvement in various aspects of the ATE program, and additional data collection when appropriate.

The papers went through several review and revision processes to ensure validity and interpretability of the findings. Outlines were submitted and reviewed by the WMU evaluation project staff and NSF. Two reviewers – generally one external expert and one of a set of three cross-paper reviewers – reviewed the first drafts. Each cross-paper reviewer read several papers. These reviewers, the authors, NSF, and the WMU evaluation project staff then met to discuss the papers, which were revised based on this feedback.

Each paper is unique. That is, each paper is based on a direction chosen by the author to best illuminate the paper's issue and hence reflects the author's personal opinion grounded by data (e.g. survey and site visit data). As such, these papers are an eclectic set, varying in tone, content, length, and the amount of judgment and recommendations. However, all are designed to provide information for ATE program illumination and improvement.

Seven of the nine papers are described in this monograph in Chapters 2–8. The papers on advisory committees and evaluation were provided separately to NSF, per its request. The major audience for these seven papers is the ATE program staff at NSF. Secondary audiences are current ATE *project* staff members and those who aspire to obtain a grant from the ATE program. Science, mathematics, and technology educators may also find these papers useful as they consider ways in which to improve their own instructional programs.

SYNOPSES OF THE ISSUE PAPERS

The following are brief descriptions of the seven issue papers (Chapters 2–8). For each, a summary of major recommendations is also included.

Collaboration

In Chapter 2, Lester Reed gives an explicit definition of collaboration in contrast to the wide range of interpretations of this issue that exist in the ATE *projects*. He provides a sample of the survey data about collaboration. He raises questions about how to interpret the survey data and uses the site visit data to help clarify and illuminate the possibilities (e.g. What accounts for the large number of collaborations per project? Do we need to have categories of collaborations?).

Reed has three major recommendations for enhancing collaborations:

- Have clear expectations. *Project* proposals should clearly identify types of membership, duration, and expected outcomes for collaborations.
- Facilitate collaboration at many levels. ATE should continue to encourage statewide, regional, or national consortia as part of the collaboration model for large *project* awards. ATE should also continue actively facilitating center-to-project and project-to-project collaboration.
- Provide guidelines. Provide guidance on "best practices" for sustaining collaborations beyond *project* funding and systematically gather data concerning collaboration sustainability.

Dissemination

Dissemination is a pervasive issue throughout the ATE program, and in Chapter 3, Thomas Owens carefully articulates the role it plays. He provides several examples of non-ATE approaches to dissemination by way of exploring potential best practices. He concentrates on the role of dissemination in centers and provides interesting additional data from a sample of ATE center directors and Web sites. He also provides some insights into the problem of how to evaluate dissemination and explores the possibility of an e-commerce approach, a proposed new paradigm.

This new paradigm may be especially appropriate for dissemination by ATE centers and includes two concepts recently borrowed from e-commerce – *integrated solutions providers* and *customer relationship management*. Combining ideas from these two concepts with the roles that centers have been playing in dissemination leads to a newer definition of dissemination – "the process of knowing your clients and systematically providing them, either directly or in partnership with other organizations, with knowledge, strategies, products and support that can enable them to better solve their problems and enhance their delivery of effective technical education." The benefits and dangers of borrowing from a business model to examine educational dissemination are addressed.

Owens provides some recommendations. Optimizing dissemination may be achieved through strengthening proposal expectations by adapting models developed for other similar programs and developing and sharing dissemination ideas, including strategies for sharing practices and evaluating impact (e.g. discussions at annual PI meeting, Web site postings).

Materials Development

Gloria Rogers' treatment of Materials Development in Chapter 4, highlights a process for producing high quality, validated materials. She demonstrates that the development of instructional and other materials is ubiquitous in the ATE program and is an issue that needs more attention. Suggestions are made that could guide this attention both in terms of NSF and of individual *projects*. This paper dovetails with the paper on Dissemination in that whatever is produced needs to be disseminated, whether only to colleagues in a home institution and/or commercially on a national basis.

Her suggestions for improving materials include the following:

- Providing resources that identify best practices in educational materials development, including NSF-sponsored workshops.
- Improving review criteria. Submitted proposals should reflect quality development processes and realistic budgets and timelines.
- Improving reporting. The current reporting process of funded *projects* should be revised to include reports on the materials development processes included in the framework for best practices.

Program Improvement

The Program Improvement chapter (Chapter 5) covers a wide range of activities and institutional levels. Lester Reed focuses this broad topic by showing examples of program improvement activities in relation to an explicit definition of program improvement. In addition to discussing the survey and site visit data, he reports on information he gathered from a sample of community college technician education program documents, which list program requirements, etc. He also addresses the interrelatedness of evaluation and materials development within the program improvement area. For example, he points out that evaluation is a necessary means to know if improvement has occurred.

Reed had four major recommendations for enhancing program improvement efforts:

- Replicate successful articulation strategies.
- Use advanced pedagogical approaches.
- Emphasize field-testing and encourage the involvement of business and industry in pilot and field-testing of programs.
- Gather program improvement production data and analyze these data in relation to U.S. requirements for skilled technicians.

Professional Development

The Professional Development chapter (Chapter 6) by Norman Gold and Karen Powe addresses the critical issue of how to provide educational resources and opportunities for instructors who are using computers in their classrooms. The authors provide interesting examples of best practices and a checklist of important components of professional development programs. They also discuss the ATE program in relation to these components and find it is doing well, although more attention to some components is necessary.

To optimize professional development opportunities, according to Gold and Powe, administrators at projects' institutions need to keep in mind:

- Increasing resources. Give faculty members time to attend courses and/or reimburse them for their time. Provide sufficient incentives to retain instructors after they have acquired upgraded skills.
- Staying current. Keep colleges and faculty current in rapidly changing technological areas and provide up-to-date equipment for faculty and staff members.

Recruitment and Retention

Given that the ultimate goal of ATE is to produce more and better prepared technicians, recruitment and retention issues are of primary importance. Arlen Gullickson and Gloria Tressler provide a three-part framework for examining recruitment and retention issues of information, preparation, and support in Chapter 7. This framework is used to synthesize the data from the survey and site visits. Descriptive vignettes illuminate the processes and provide insights into existing practices. In addition, the authors provide a succinct statement of what NSF should consider including under the heading of recruitment and retention.

According to the authors, enhancing recruitment and retention can be accomplished through three steps:

- Building a comprehensive program. Encourage *projects* to develop recruitment and retention as a system of interrelated activities so that all persons (potential students, future employers, instructors, and support staff) have a thorough understanding of the program and its integral parts.
- Revising guidelines to encourage strong recruitment and retention programs.
- Collecting data and using it to improve retention and recruitment programs.

Sustainability

Because sustainability is such a complex issue, the paper by Frances Lawrenz and Nanette Keiser (Chapter 8) considers it from a variety of perspectives. In a simple sense, sustainability for the ATE program could mean continuation of whatever activities had been supported by the NSF grant, including institutionalization activities. This is consistent with the definition of sustainability given by the Community College Research Center (CCRC) in its study of the ATE program. CCRC defined sustainability as "The state where the major activities involved in the ATE program continue even after the grant expires." The existing literature on how to sustain organizational change is synthesized into a checklist, which serves as a backdrop for considering the possibilities and desirability for sustainability within the ATE program.

It appears that ATE *projects* engage in many activities that will be sustained, including institutionalization. Many factors support this conclusion including diverse resources brought to bear on the *projects* and distributed power. Wide participation, clear purposes, and knowledge are often evident. Challenges to sustainability include information that is not always available or used to reward effort and inconsistent marketing and promotion.

Lawrenz and Keiser concluded that maximizing sustainability may result from these activities:

- Clarifying ATE program goals for sustainability and determining how to help *projects* achieve them.
- Using a sustainability plan. Integrate sustainability strategies into ATE *project* work from the beginning.
- Collecting sustainability data. Place more attention on data collection to identify *project* components that should be sustained, to learn how to improve components, to provide information upon which to base rewards, and to convince others of the worth of the components.

THE ATE PROGRAM

Origins of the ATE Program

In the early 1990s, there was a national interest in a balanced approach to developing and using technology[8] to meet the nation's educational and workforce needs. The importance of such initiatives is clearly developed and described in *Technology for All Americans: A Rationale and Structure for the Study of Technology* (1996) and *Gaining the Competitive Edge: Critical Issues in Science and Engineering Technician Education* (1993). As those documents indicate, this country has a critical need for trained, professional technicians with unique skills in technology and technological systems. These persons must be educated to serve emerging needs of business and industry and must be able to work on applications that build on theoretical understandings. On October 23, 1992, Congress passed the *Scientific and Advanced-Technology Act of 1992* (PL 102–476), which called for NSF to establish a national program to improve the education for technicians in advanced technology fields. This became the ATE program.

This Act was intended to serve the ultimate goal of improving the competitiveness of the U.S. in international trade by increasing the productivity of the nation's industries, which in turn was to be accomplished by increasing the pool of skilled technicians in *strategic* advanced-technology fields. It is noteworthy that Congress emphasized the role of two-year colleges for this program. As House Report 102–508, p. 4, states, "Two-year colleges are a major contributor to higher education and have become the largest pipeline to postsecondary education in the United States. In 1990, 1,350 two-year colleges enrolled approximately 5 million students, representing 43% of all undergraduate students and constituting 40% of all institutions of higher education. Approximately

30% of students enrolled in two-year colleges transfer to four-year colleges and universities."

Additionally, Congress sought to define and delimit what it included in the realm of advanced technology. As stated in the bill, "the term 'advanced technology' includes advanced technical activities such as the modernization, miniaturization, integration, and computerization of electronic, hydraulic, pneumatic, laser, nuclear, chemical, telecommunication, fiber optic, robotic, and other technological applications to enhance productivity improvements in manufacturing, communication, transportation, commercial, and similar economic and national security activities."

This increased pool of technicians was to be accomplished through a direct focus on and funding of targeted educational programs. Congress identified four purposes to be served by the Act and the ATE program: (1) improve science and technical education at associate-degree-granting colleges; (2) improve secondary and postsecondary school curricula in mathematics and science; (3) improve the educational opportunities of postsecondary students by creating comprehensive articulation agreements and planning between two-year and four-year institutions; and (4) promote outreach to secondary schools to improve mathematics and science instruction.

ATE, compared with programs like Carl Perkins and many NSF programs, is not a large program. Although it has grown from $14.5 M to $40 M, it cannot by itself create all the technicians needed by the country.

ATE Program Purposes

NSF initiated the ATE program to address the Congressional mandate. The ATE program was created in the Education and Human Resources Directorate (EHR) and is co-managed by two Divisions: the Division of Undergraduate Education (DUE) and the Elementary, Secondary, and Informal Education Division (ESIE). The overall purpose of ATE is to provide a better-educated workforce. To accomplish this, the goals of the ATE program are to: (1) produce more science and engineering technicians to meet workforce demands; and (2) improve the technical skills and the general science, technology, engineering, and mathematics (STEM) preparation of these technicians and the educators who prepare them. Additional information about the ATE program is available at [http://www.ehr.nsf.gov/ehr/due/programs/ate/].

ATE focuses its funding efforts at the community college level in order to strengthen and expand the scientific and technical education and training capabilities of associate-degree-granting colleges. ATE set priorities for what types of work would be supported and how it would allocate funding. The ATE program has four objectives to accompany its goals and overall purpose:

- Develop model instructional programs in advanced-technology fields (program improvement).
- Provide professional development for faculty and instructors in advanced technology fields (professional development).
- Establish innovative partnership arrangements (collaboration) with associate-degree-granting colleges, secondary schools, colleges/universities, businesses, industries, and other appropriate public and private sector entities that need skilled technicians in their workforces. The partnerships with industry are to make certain that the education provided by ATE is consistent with industry's needs.
- Develop and disseminate instructional materials (materials development and dissemination).

ATE Projects and Centers

Grant awards are made in two categories, centers and projects, with centers individually receiving substantially more funds and having a broader scope. Each center or project uses these funds to facilitate the development of technicians. While a project tends to focus on only one or two of the above objectives, centers typically address all or most of the objectives. Centers always receive funding for multiple years, serve as model programs for other institutions and organizations, and disseminate information to a region (e.g. several states or the nation as a whole).

Since its inception in 1994, more than 400 grants have been awarded. While most have gone to two-year colleges, ATE also offers outreach to secondary and baccalaureate institutions. About 200 *projects* are currently active across the United States. ATE supports education in a broad range of technology fields including agricultural technology, biotechnology, chemical technology, civil and construction technology, computer and information technology, electronics, environmental technology, geographic information systems, manufacturing and engineering technology, marine technology, multimedia technology, telecommunications, and transportation technology. ATE also supports education in the basic sciences needed to understand the technologies (e.g. physics, chemistry, and mathematics).

ATE program activities include the adaptation of exemplary educational materials, courses, and curricula in new educational settings; the design and implementation of new educational materials, courses, laboratories, and curricula; the preparation and professional development of college faculty and secondary school teachers; internships and field experiences for students, faculty, and teachers; and the broad dissemination of exemplary educational materials and pedagogical strategies that have been developed through funded ATE projects. Activities may have either a national or a regional focus, but not a purely local one (*Advanced Technological Education [ATE] Program Solicitation, NSF 01–52*).

THE WMU EVALUATION PROJECT

Purpose

NSF funded the evaluation project at Western Michigan University's Evaluation Center to assess the impact and effectiveness of the ATE program in 1999. This evaluation addresses four basic questions important to ATE and its stakeholders: (1) To what degree is the program achieving its goals? (2) Is it making an impact, reaching the individuals and groups intended? (3) How effective is it when it reaches its constituents? (4) Are there ways the program can be significantly improved? In its first two plus years of work, the project has collected and provided evaluative information and judgments in all these areas.

Methodology

The project has employed two primary mechanisms to gather data: a Web-based survey and site visits. The Web-based survey provides an effective way to gather impact and effectiveness data to serve general accountability purposes. The site visits have been especially helpful in putting a "human face" on the *project* efforts and the institutions within which the *projects* conduct their work. The site visits have also greatly facilitated understanding of the *projects* and helped clarify and improve findings and recommendations developed from *project* documents and survey results.

NOTES

1. As described in detail on page 8, NSF funds projects and centers. The convention *projects*, in italics, will be used to denote projects and centers.
2. Joey Suing, graduate of Pasadena City College's Academy for Creative Technologies. Curriculum developed for this program through funding from the ATE program. Ashlock, T., and Wright, S. (2001). *Students at The Learning Edge: Advanced Technological Education Programs at Community Colleges*, p. 4.

3. One of the students interviewed during 13 site visits.

4. American Association of Community Colleges (2000). *The Learning Edge: Advanced Technological Education Programs at Community Colleges.*

5. *Ibid.*

6. A Web-based survey of the active ATE *projects* was conducted in May 2000 ($n = 113$) and again in February 2001 ($n = 81$). The purpose of the Web-based survey was to better understand the nature of the ATE *projects* and to begin to address the effectiveness of these grants. The survey consisted of nine sections. In 2001, only minor changes were made to the survey instrument used in 2000. The reports detailing the findings and recommendations from these surveys may be found at http://www.ate.wmich.edu (To access the reports, click on Evaluation Products at this site and then click on the appropriate heading for the reports [i.e. *Status Report 2*for Survey 2000; *Survey 2001: The Status of ATE Projects and Centers*for Survey 2001]).

7. The ATE evaluation project conducted 13 site visits (3 centers and 10 projects) from September 2000 through February 2001. Judicious mixes of knowledgeable evaluators and/or content experts conducted the visits at these sites, which were deemed to be representative of the ATE program. Sites were selected using a purposive sampling technique based on recommendations from NSF, thus ensuring a cross section of *project* types in the ATE program. To minimize the burden on the sites, site visits were usually one full day in length for projects and two full days for centers. Each site received a report that was for its use only. Prior to the finalization of reports, sites were provided an opportunity to comment on the accuracy and perceptions of the draft reports. All 13 final site visit reports were provided to the issue paper authors who then aggregated the findings relevant to their topics.

8. "Technology" is defined as the branch of knowledge that deals with the creation and use of technical means and their interrelation with life, society, and the environment, drawing upon such subjects as industrial arts, engineering, applied science, and pure science.

COLLABORATION EFFORTS: APPROACHES, USE, AND EFFECTIVENESS OF COLLABORATION

Lester W. Reed Jr.

THE ORIGINS OF ATE COLLABORATIONS

The ATE program is NSF's response to the *Scientific and Advanced – Technology Act of 1992 (PL 102–476)*. In creating the Act, Congress stated a belief that the issue of under-preparedness could best be addressed by *collaboration among the nation's associate-degree-granting colleges and private industry* and concluded that NSF's role in *stimulating partnerships between educational institutions and industry* made an enlarged role in scientific and technician education particularly appropriate.[1]

By establishing key goals of *collaboration* and *partnerships*, Congress is clearly calling for a joint effort involving education deliverers and industry in NSF-ATE funded programs. Equally apparent is the focus on associate-degree-granting colleges as the initiator of these supportive arrangements. ATE award guidance states that *projects* should include "two-year colleges leader-ship roles."[2] This stipulation remains consistent in award guidance for 1997 through 2002.[3]

This paper addresses the concept and results of *collaboration* and partnerships in funded ATE centers and projects. Centers are a major effort by a funded

Evaluating the Upgrading of Technical Courses at Two-Year Colleges:
NSF'S Advanced Technological Education Program
Advances in Program Evaluation, Volume 9, 13–33
Published by Elsevier Ltd.
ISSN: 1474-7863/doi:10.1016/S1474-7863(04)09002-7

entity spanning multiple years, whereas projects are more limited in their objectives.

DEFINING COLLABORATION

An operational working definition of *collaboration* is needed as a starting place in reviewing the degree of success ATE-funded *projects* are having in meeting the goal of collaboration and partnerships set by Congress. Three primary sources can contribute to this definition: existing literature on the topic,[4] the definition formed by a review of NSF-ATE program guidance and solicitation documents, and the definition developed by The Evaluation Center as part of its evaluation of the ATE program.[5] Although the term *collaboration* is primarily used in this paper, it includes the concept of *partnerships* that is commonly used in educational literature when referring to mutually beneficial and supportive relationships.[6]

The review of literature focused on collaborations in an educational setting. Other than dictionary definitions, educational literature generally describes the need for and intended results of collaboration and/or partnerships but, with some exception, rarely discusses the dynamics of such a relationship.

A somewhat typical approach to collaboration in literature is to list the expected behaviors of specific partners. An example is the list provided for partnerships to the improving science, technology, engineering, and mathematics (STEM) education. This list contains 22 actions such as *Provide funding for technology* and *Provide state-of-the-art equipment to local colleges and universities*.[7]

The *Scientific and Advanced Technology Act of 1992* also assumes that the concept of collaboration between private industry and other entities and the community colleges[8] is clearly understood. This seems to be a safe assumption for technical education programs. A review of literature addressing community college educational programs points to the fact that collaborative relationships between community colleges and external and internal entities are essential in career or technical programs. Also evident is that a link between the community college and employers has been a staple of workforce development programs since they were first introduced in the 1950s.[9]

Typically, community colleges report that the nature of technical programs requires partnerships with business and industry. For example, Madison Area Technical College (MATC), Wisconsin, reports, "The concept of partnerships evokes a variety of models used by academic institutions to build and maintain interactions with business and industry."[10] In describing the technical program partnerships, MATC states, "The curriculum and equipment needs of the programs were defined through input of business and industry partners."[11] However, there is no discussion of the components or dynamics of the collaboration engendered by these college/business partnerships.

ATE guidelines point to the need for *projects* to form formal and sustained multiple cooperative arrangements supporting a variety of outcomes. For example, ATE materials refer to *projects* "establishing partnerships among high schools, businesses, government agencies, and professional societies in order to respond to educational needs of the workforce."[12] ATE documents also include expectations that its programs "promote exemplary improvement in technical education at national and regional levels by supporting – particularly in two-year colleges and secondary schools – the design and implementation of new curricula, courses, laboratories, instructional materials, opportunities for faculty and teacher development, academic support for students, and formal cooperative arrangements among educational institutions and partners from business, industry, and government." [13] The program announcement further states that these alliances should exist "both during the *project* and on an ongoing basis after its completion."[14] Although clearly an expectation, guidance as to how to arrange and nurture the referenced "partnerships," "cooperative arrangements," and "alliances" is not provided by ATE.

This theme of multiple collaborative arrangements is further expanded in *The Learning Edge*[15] that states, "*projects* should be built on alliances of associate degree granting institutions with four-year colleges and universities, secondary schools, business, industry, and government."

A final source of input into this paper's operational working definition of collaboration is the definition offered by The Evaluation Center in its review of components driving successful accomplishment of funded *projects*. The Center's definition states "*collaboration is the relationship of projects and centers with businesses, industries, educational institutions, and other organizations to achieve project/center objectives*."[16]

Collaboration Defined

The definitional construct used in this paper is: Collaboration is a sustained formal partnership fostered by ATE projects and centers between K-12 schools, community colleges, four-year colleges and universities, businesses, government agencies, and professional societies in order to respond to the educational needs of the workforce by facilitating the achievement of the project's/center's objectives and which results in mutual benefit to all participants.

Collaborations in Technical Education

ATE is essentially engaged in technical education with a preponderance of the effort focused at the community college level. However, even without the influence of ATE funding, all community college technical programs are, by necessity, engaged in collaborations. The nature of the curriculum demands association with product users – businesses and industries that hire graduates. Such associations may be from ad hoc gatherings for a limited purpose (such as defining standards) to standing committees that persist through the life of the program. For example, technical programs customarily have advisory committees that include business and industry representatives. In fact, such "lay advisory committees" for technical degree programs is a community college system mandated requirement in almost all states. Dependent on local requirements, these committees meet to assist programs from as frequently as quarterly to only once annually. Career and vocational programs at the secondary school level also use similar advisory committees.

Consortia of two-year colleges exist in almost all systems to provide coordinated educational programs. And, for programs that have articulation with four-year programs as a goal, there are collaborative arrangements with the senior institutions. Often, program directors and technical faculty are members of institutional teams addressing recruitment and student development services. These same individuals often work with K-12 schools to foster approaches that lead students to enter their technical programs.

What is different is that ATE *projects* are expected to expand and raise collaborations to a higher level of success. Collaboration, particularly with business/industry, K-12 programs, and other two-and four-year educational institutions, is a major pillar on which ATE material and program development rest. Collaboration for ATE *projects* is not a secondary effort, but a major supporting activity vital to quality technical education. ATE-assisted technical education efforts are expected to be successful, due in great part to the use of effective collaboration.

Key Collaborative Areas

Technical workforce development efforts that require sustained collaborations with business and industry include areas such as skill/standards development, curriculum review, providing work-based education experiences, and program pilot and field-testing. Pilot and field-testing of curriculum are often thought of only in the context of activities conducted by members of the academic community. However, business and industry participation in these activities is essential in technical education programs. The issue for business and industry is the verification of curriculum meeting the intended outcomes of developing workplace knowledge and skills. The academic community engaged in technical education can design and collect pilot or field-test data that reflect on the efficacy of the educational process. But the final question of "Does it produce advanced technicians for the workforce?" is one that must be answered by those who hire the technicians.

A special benefit of meaningful collaboration between business and industry and *projects* is credibility. If the employers are part of the process of designing, providing, and evaluating workforce programs (or materials used in such efforts), they are more likely to hire the individuals educated in the ATE *projects*. The business and industry members of a standards development team may not have responsibility for hiring; however, the fact that their contribution is part of the basis of a technical program provides the academic institution leverage in the placement of program completers. Assuming that the technicians emerging from the improved programs are better prepared, then the productivity of the American workforce will be enhanced. The transferability and dissemination of products from ATE *projects* will also be enhanced if they are supported by the businesses and industries employing individuals who were educated using ATE-generated products. Also, these collaborations often serve to expand the resources of the *projects*, particularly in areas of equipment, software, and other industry-specific educational needs.

Two other areas that are collaboration-dependent are the articulation of programs and the K-12/community college connection. Articulation (the movement of a student's educational experience between educational entities) depends on the willingness of the receiving institutions to accept the transferring student's educational competencies as equivalent to those provided by the receiving institution. Continuing collaboration on the content/outcomes of courses between secondary to associate degree and associate to baccalaureate degree institutions is essential for viable articulation agreements.

The K-12/community college connection is another area where sustained dialog and mutual effort is essential to ATE's success. The major areas of this effort are teacher enhancement and recruiting of students. Improvement in student readiness in science, mathematics, engineering, and technology (SMET) is a major ATE objective. Community college and K-12 collaboration on increasing teacher development, particularly in technology, provides a "win-win" situation for the K-12 school system and the community college. The result of this effort is to provide better-prepared students for advanced technician programs at the community college.

STATE OF ATE PROJECTS USE OF COLLABORATION

This section of the paper is oriented to the current state of collaboration in funded ATE *projects*. The sources for the data presented are listed below:

- A detailed review of The Evaluation Center's report of *Findings from a Survey of ATE Projects and Centers* (a Year 2000 report and draft of the Year 2001 report).[17]
- A review and comparison of factors discussed in site visit reports[18] from visits to selected ATE *projects*.

As a departure for measuring ATE *project* collaboration, The Evaluation Center's report of *Findings From a Survey of ATE Projects and Centers* provides a measure of the *projects'* self-reported collaboration arrangements. In the Executive Summary of the Year 2000 report,[19] Finding 2 states, "ATE *projects* have established a large number of collaborative arrangements. The collaborations serve multiple purposes and provide monetary support as well as other kinds of assistance for materials development, academic programs, and professional development efforts."[20] A similar finding was presented based on the 2001 survey data. The finding goes on to state that overall *project* efforts have yielded more than 15,000 collaborations (over 13,000 in 2001 survey). However, the nature of aggregated findings limit analysis in terms of *projects* achieving meaningful relationships in line with Congressional and NSF ATE expectations.

The 2000 and 2001 survey reports do caution that *projects* might have reported a single collaboration in multiple categories, resulting in the large totals. After applying an adjustment for the maximum in the data, the report authors estimate that in 2001 there would still be more than 900 collaborations – nearly 16 per *project*. Also, all 13 *projects* at which site visits were conducted had numerous collaborative arrangements. Regardless of the exact number of collaborative arrangements, ATE *projects* clearly are collaborating with others as they pursue their goals.

Collaborations reported in the survey were with a variety of institutions and organizations (e.g. business/industry, secondary education, associate and baccalaureate degree institutions, and professional associations) and for multiple purposes (e.g. professional development, materials development, and advisory). Of the reported collaborations, *projects* identified direct or in-kind funding as the nature of the relationship in terms of dollars received as a result of the collaboration. Direct contributions of money from non-NSF sources remained relatively constant (around \$12–\$14 million) in both survey years. In each year, *projects* reported leveraging NSF's funds with additional monetary and in-kind contributions from non-NSF sources. For every dollar provided by NSF for the duration of the *projects'* grant periods, the *projects* reported increasing their working resources for the ATE program by 50 cents in 2000 and by 80 cents in 2001.[21] ATE *projects* are clearly successful in leveraging their grant dollars to attract significant additional resources through the collaborative process.

The survey categorized collaborations by four types of organizations and purpose. The greatest numbers of collaborations were with business/industry (nearly 85% for projects and 100% for centers in 2001). Collaboration rates with other organizations identified in the survey ranged from about 50 to 65% for *projects* and 75–100% for centers. The one exception was the catchall category of "other." With the comprehensive list of identified organizational types in the survey, a limited number of "collaboration with others" seems appropriate. Considering the emphasis on technical education, this distribution reported in the survey is in line with the author's expectations and reflects a comprehensive approach to collaboration by ATE *projects*.

In terms of the collaborative purpose, data are presented as a percentage of the *projects* having collaborations in areas of general support (advice, shared equipment, etc.); materials development (developing standards, pilot and/or field-testing materials, etc.); professional development (e.g. providing knowledge of industry needs, developing faculty knowledge and skills, etc.); and program improvement/academic programs (e.g. work-based instruction, student recruitment, student understanding of industry requirements, etc.). These data show a relatively high percentage of collaborative support from business/industry and educational institutions in all categories. This distribution reflects ATE *project* engagement in collaborations over a large landscape of potential purposes.

In site visit reports, the identified collaborations were similar to that reported in the survey data and were primarily with educational entities (K-12 schools, community colleges, and baccalaureate degree colleges) and businesses and industries. In most cases, the *projects* developing instructional materials or creating improved technician programs were housed at community colleges. The type and scope of collaborations in the associate degree colleges varied in length of the relationship, purpose of the collaboration, and degree of involvement by participants.

A significant group of collaborations cited in site visit reports were short-term (even single encounter) relationships. Meetings with local business/industry representatives to identify workplace competencies most often reflected this limited, one-time characteristic. Regardless of their length, business/industry collaborations were, however, critical to the success of the *project* by providing needed advice; access to equipment; and in some instances, funding. The reports provided some examples of longer term collaborations with business/industry, such as advisory committees and developing and providing student internships that also were essential to improved technical education. The site visit interviews with collaborative business/industry representatives reflected their respect for the *projects'* efforts, and industry involvement clearly increased the credibility of the *projects'* product(s).

An interesting data element from the survey is the reported low participation of business/industry in pilot and field-testing of materials. Three of 11 centers and 6 of 46 projects[22] reported pilot/field-testing as the nature of their collaborations with business/industry. Most centers and nearly half of the projects reported support for pilot and/or field-testing by educational institutions. This may signal a need to strengthen this relationship between business/industry and *projects* in verifying curriculum products.

The ATE *projects* reported that the quality and productivity of their collaborations ranged from satisfactory to excellent. Centers reported a lower degree of quality and productivity than did projects. Since the number of collaborations managed by each center is larger than those of each project, the greater number of relationships that are less productive or of lesser quality at centers could reasonably be anticipated.[23] Comments in site visit reports generally rate collaborations as meaningful and productive.

The survey collected data on factors that ATE *projects* identified as significant in enhancing or creating productive collaborations. Respondents identified factors such as mutual benefit to collaboration participants, a clear statement of goals, defined roles of partners, established communication channels, building trust, fostering group understanding of goals, and achieving a zone of comfort among team members.

Barriers to productive collaborations focused on two major areas – lack of resources and time factors. Reported resource constraints affecting collaborators are a lack of top-level school and college administrative interest or support. In general, all *project* levels cited the following items as "resource barriers": (1) misunderstanding of the "*project idea*" (or purpose or goal); (2) entrenchment – especially in higher education; (3) overstating goals in underfunded grants; and (4) competing requirements for limited resources. Lack of time by collaboration academic partners to create or review curricular materials and the inability of business partners to "miss work" were time factors cited as barriers. Although time was the most cited reason for collaboration failure, the interaction of time with other factors, such as clarity of purpose that establishes a high priority for collaborative effort, might signal that "lack of time" serves as a surrogate for "lack of interest" or other factors that need fixing to insure success. [24]

Project leaders' collaborations with K-12 school systems were generally at the secondary level and varied from providing materials, making presentations to students and, in some cases, working toward secondary to postsecondary articulations. *Project* participants also developed and presented workshops in cooperation with school districts and schools. The description of these relationships revealed that they often were heavily reliant on *projects* "providing services" and "initiating contact" with schools and not based on a concept of shared effort and purpose. The secondary/postsecondary relationship seems to be motivated by attracting students to enroll in the *project's* program. In this regard, the reported K-12-community college relationship focused on improving technician education, since the associate degree is the main vehicle for that improvement. The efforts of *projects* to increase K-12 teacher preparation for the understanding of and opportunities available in technical education fits well with the ATE initiative. However, these efforts often existed before the NSF ATE grant, but seemed to be enhanced by the infusion of grant funds. The activity level in these secondary/postsecondary collaborations after the funds are exhausted remains to be seen.

With almost no exception, community college collaborations with institutions having four- year programs in disciplines related to the *project's* educational program were for articulation. These relationships, although sometimes characterized as receiving advice on program content and/or verifying appropriate course content by senior colleges, were for the purpose of aligning course content for acceptance as part of a student's four-year program. Achieving the maximum degree possible in educational continuum "seamlessness" is a goal of the ATE program and is being well supported by the *projects*. Based on reported site visit data and discussions with *project* staff by the author during such visits, the characteristics of these collaborations:

- are individually oriented toward a specific senior college and discipline;
- persist once an agreement is reached and formally executed;
- are normally initiated and pursued by the associate degree institution (although in some reported cases, the senior institution's need for increased enrollment resulted in it being an aggressive vs. a reluctant partner);
- once the agreement is reached, is a need for periodic recontact by the *project* institution to ensure a continuing, beneficial implementation of the agreement.

In some reported collaborations, agreements on dual enrollment and shared tuition between associate and baccalaureate institutions have moved traditional articulation to a new level. In general, the associate degree institution's effort to achieve articulation of its technical programs predates ATE. However, as is the case with the secondary and postsecondary collaborations, ATE resources have had an impact. Funding from ATE to improve programs has served as a stimulus for acceptance of credits by senior colleges. In the future, pursuit of collaborations to create articulation agreements will persist even without ATE's additional funding. But, as evidenced by the reported newly negotiated articulation agreements, ATE influence in creating more rigorous technical programs at community colleges has created a more willing environment for mutually beneficial collaborations.

A collaboration between projects with similar objectives and, more likely, between centers and projects would seem natural for the ATE program. However, the site visit reports make reference to only one such collaboration.

FINDINGS BASED ON THE STATE OF COLLABORATION

Presented below are some generalized findings based on data gathered from the surveys and visits to ATE *projects*.

Effectiveness of Project Collaboration

How do the data on reported ATE project collaborations stackup against the definition offered in this paper? An overall answer would be "very good." As a reminder, the proffered definition follows:

> Collaboration is a sustained formal partnership fostered by ATE projects and centers between K-12 schools, community colleges, four-year colleges and universities, businesses, government agencies, and professional societies in order to respond to the educational needs of the workforce by facilitating the achievement of the project's/center's objectives and which results in mutual benefit to all participants.

A summary of ATE collaborations compared to the definitional elements is presented below:

- *The collaborative relationship is to be initiated, facilitated, and sustained by the ATE project.*
  All collaborations were initiated by ATE *projects*, even if they existed before ATE funding. The need to focus on an issue, in some instances, arose outside of the *project* organization, but the *project* organization assumed responsibility for creating collaborations needed to resolve the issue. Those *projects* that produced positive results were guided and managed by *project* personnel. Collaborations generally atrophied without active leadership by *projects*, and collaborations with a limited purpose also had a limited life. In particular, collaborations involving business and industry in developing workplace standards were reported to be active early in the *project's* life. However, they did not always

evolve into a continuing, mutually supportive relationship during curriculum development, testing, and implementation.

- *The purpose of the collaboration is to support objectives of the ATE project and should be formally defined by the parties concerned.*
  The trigger for all reported collaboration was the support of *project* objectives. In some cases, the linkage to NSF-ATE support of these objectives was not made clear to collaborators. In one report, the comment was made that " it appeared that NSF support was the silent or invisible partner in many of the programs described." Based on site visit reports, the foundation of the most successful and sustainable collaborations was a formal understanding of purpose, membership and expected contributions, procedures, and identification of expected results. Although not specifically discussed in these reports, it can be deduced that formalization was the exception, not the rule. Consortiums with shared grant funding were most likely to have a more formal structure.
- *Although short-lived and/or serendipitous relationships can benefit goal achievement, the collaboration envisioned by Congress and NSF-ATE is a sustained systemic effort during and after ATE funding.*
  The strongest collaborations were characterized by the routine and regular meetings of the parties involved. These contacts appear necessary to sustain enthusiasm and progress toward stated goals. Although it could be assumed that collaborations for articulation with four-year programs and secondary school contacts associated with recruiting would continue, the postgrant level of activity is not known. This is due in part to the evaluation structure that examined only currently funded *projects*. Anecdotal data in several site visit reports point to the fact that collaborative efforts between resource partners did not remain strong after *projects* ended funding for the activity.
- *Participants in collaboration include all who have a stake in the outcome of creating a world-class workforce. The following major players are included:*
  - *K-12 schools as feeders into community colleges*
    This clearly was one of the major roles undertaken by *project* leaders, particularly in the community-college-based *projects*.
  - *Community colleges that produce the advanced technicians*
    Consortiums of community colleges are a major feature of many *projects*. These national, regional, and statewide consortiums based on supportive associate degree programs appear to be the strongest component in the ATE *project* matrix.
  - *Four-year colleges that provide advanced educational opportunities for associate degree technicians*
    Associate degree colleges invariably pursue program articulation with baccalaureate programs for their technical degrees. The support of ATE appears to increase the success of achieving articulation.

- *Government and governance entities that control and facilitate the educational processes*
  Statewide initiatives tend to create coalitions that involve state agencies in the task of approving and creating acceptance for ATE *projects'* efforts. This support appears to be beneficial in terms of moving toward the goal of a wider acceptance of the *project's* products by state system community colleges.
- *Private and public businesses and industries that employ the technicians*
  All ATE *projects* that took part in the site visits had multiple collaborative arrangements with employers in the private and public sectors. Based on both survey data and site visit reports, the relationships seemed to be directed toward limited purposes (e.g. standards development, development of student work-based educational experiences, equipment and/or funding support, etc.). Many collaborations were of relatively short duration, such as calling together representatives to assist in identifying workplace competencies. There were longer relationships, particularly in terms of lay advisory committees for specific technical programs. These relationships exist as long as the educational program exists regardless of supplemental ATE funding. Implicit in the site visit reports is the fact that wider based *projects* (national, regional, or statewide) have a more formal enduring relationship with business/industry. Based on site visit and survey data, business/industry was not significantly involved in verifying the efficacy of *project* materials by participating in pilot or field-testing.
- *Professional organizations that support elements of educational improvement*
  Program accreditation by professional organizations was sought and gained by some of the ATE programs. Based on reported data, most programs in information technology sought and received industry certification.

- *Partners in collaboration must have a recognizable stake in the effort's outcomes. Collaborative results that meet partner needs should be clearly identified.*

Collaborations clearly support the goals of the ATE *projects*. There are numerous examples where the mutually beneficial relationship between partners was clearly understood. This is particularly true when business and industry had a critical need for skilled technicians and in associate degree/baccalaureate degree articulations. Reports also indicate that both parties in K-12/community college relationships that provide teacher professional development actively embrace *project* goals. However, it is not clear from the available data that all partners in collaborations had their stakes in the effort clearly identified. In one reported relationship between a large urban and a smaller rural community college, the larger college appeared to view the collaboration as of little or no benefit and only marginally pursued the stated common goals. Also, the goal of instituting changes to K-12 school curriculum (both in increasing math/science rigor and integrating technology topics) appears to be less productive for a variety of reasons that are beyond the control of the ATE *project* leaders.

Collaboration Models for Expanding Project Impact

The primary focus of the ATE program is the improvement of the U.S. workforce, not just a local workforce. This presents a challenge since, with some exceptions, the community college is the ATE grantee for projects, and community colleges are just that – community based. This community orientation may restrict the reach of the community college to its geographical service area.[25] The issue is "How do these traditionally locally focused institutions expand their influence beyond their traditional geographical and constituent boundaries?" Collaboration with other community colleges is the best answer. Other activities, such as materials clearinghouses, conferences, workshops, and presentations at meetings, can contribute. But, based on reported data, to expand the ATE effort past the local college, sustained personal college-to-college relationships work the best. Community colleges working together on similar technical programs provide a vehicle for focused adaptation of materials and establishes a mutual support system. During site visits, projects reported that, by working in unison with other colleges, their efforts had a greater impact than if products were simply "made available."

The site visit reports provide several examples that are clearly models of this type of collaboration.[26] A description of the characteristics of these success models is presented below.[27]

Model 1 – The National or Regional Decentralized Consortium
This arrangement creates a focal hub point that has overall responsibility for the *project's* activities and is the contact point for administrative and management functions. The focal point, in turn, develops a national or regional network of semiautonomous community colleges that serve a distinct geographical area and are responsible for facilitating achievement of the mutually agreed upon consortium goals in their area. A formal structure that fosters communication, sharing of resources, developing and sharing materials, and assessing progress is created. The structure provides for scheduled progress review meetings of the consortium focal point and decentralized subproject leaders. In this model, the decentralized subproject entity (usually a single community college) forms its own collaborative arrangements with area schools, business/industry, and senior colleges or community colleges. The decentralized subproject also has specific responsibility for materials development and program improvement in one of

the consortium program areas. The central focal point may or may not have such a responsibility.

This model capitalizes on the unique relationship community colleges have with their communities. The local colleges (assuming a positive reputation) can marshal resources in their area and achieve the workforce improvements in that local area that would not be possible for an outside college. These improvements have a central core of competencies in line with ATE objectives and expand the influence of ATE-supported efforts to a region, not just a single community. The key to success is ensuring mutual interest in achieving the goals and a willingness among partners to work cooperatively with others. A major advantage of this model is that turf battles are avoided, since as each subproject operates in its normal service area and retains its own program and managerial autonomy. A note of caution: In reviewing reports on the effectiveness of the *projects* using this model, it was evident that sufficient resources must be allocated to manage the national/regional consortium or an excess degree of independence (i.e. lack of oversight and monitoring by the focal point) of partners can reduce the effectiveness of a *project* in meeting overall goals.

Model 2 – Statewide Consortia

In this model, the state system of community colleges is involved as a player. The funded entity could be the state governance agency for community colleges or one of the colleges in the system. The focus is to create or improve technical programs to meet statewide economic development goals through workforce development. State system colleges serve as partners in the development effort including standards and curriculum development and faculty professional development. The business and industry participation is normally a mix of statewide and local collaborations, and the state economic development agency is normally a significant contributor toward enlisting business/industry support. The state department of education often provides support through its ties with local school districts. However, the local community colleges still must create local school system interest in a mutually supporting relationship. Depending on the community college system's structure, several geographically strategic community colleges may serve as regional coordinators and as part of a *project's* management team and receive funding support from the grant. Once developed, products are made available to all colleges. Articulation agreements with senior colleges normally remain a college-by-college responsibility. However, as a statewide effort, there seems to be potential for statewide articulation, although none was reported from the sites visited.

A variation of this approach is for a limited consortium of a group of state community colleges. These may be formed around aligned geographic service areas or common business or industry workforce needs. In the limited consortium, there is less direct involvement of state agencies than in the statewide approach and more reliance on perceived mutual benefits resulting from the collaboration. As in most consortiums, one of the players is required to be the focal point for the ATE effort.[28]

In reviewing the site visit reports describing statewide *projects*, it was clear that the strength of the community college system is a critical factor. State community college systems that have structured procedures and mechanisms for developing and approving curriculum materials are the most successful. Under such an established protocol, uniform statewide student competencies, curriculum materials, and course syllabi are readily accepted without turf wars that arise when each institution has its own approach. With an in-place system, the energy normally expended on reaching procedural consensus can be directed to improved product generation. One final note concerning the statewide model, a systemwide effort can often bring added support from the state's executive and legislature branches. This was evident in the report of one such project.

Model 3 – Central Product Development with Nationwide Users

In this approach, the curriculum product development is centralized and managed by the *project*. Product development can be done in-house or by compensated development teams. Liaison with industry professional societies and/or user groups is often established to stimulate cross- industry support vs. dealing with one company. This nationwide collaboration with a specific industry provides product creditability. The products that best fit the centralized development are instructional modules that can be adapted by multiple users in a variety of settings. The central *project* leadership is responsible for attracting "customers" for the product. Although the customers are characterized as "partners," their collaboration is (at least in the reported project) limited to buying and using the product. The strength of this model is the potential distribution of the products nationwide in support of a major industry. However, interesting users who have had little or no say in the product's development is a challenge. The credibility of an industrywide-supported product can decrease the impact of this challenge. Also, this model can result in sizable central *project* staffs that may not be sustainable once grant funding has ended.

An example of creating a nationally available product is the use of professional societies or similar organizations with a specific interest in improving technical education. In the reported data concerning such an arrangement, the collaborations formed involved various industries as product advisors and materials development sites. Additionally, a national network of education professionals served to pilot test and critique the materials during development. A professional publisher was the catalyst for producing the finished product and its distribution. Plans were in place for expanding and updating the materials using the initial collaborative process. In this approach, the ongoing collaborator was identified as the publisher; however, this was clearly a commercial arrangement and not one growing out of workforce development concerns of the publishing company and would more correctly be characterized as a dissemination approach.

Site visit reports provided data on several single college efforts under the ATE program. In analyzing these data, it was evident that the colleges involved did produce improved technician education programs with an expanded impact. They essentially used the types of collaborative arrangements described above. However, these arrangements, particularly with business and industry, were limited in scope (often one or two meetings with representatives of local small businesses). This type of collaboration may serve to develop a program meeting local needs, but would seem to have a minimal chance of providing widespread adaptation by other colleges.

CONCLUSIONS

Based on those data contained in the WMU evaluation project surveys and site visits, it is clear that ATE projects/centers are using collaborations effectively. The overall positive impact of these collaborative arrangements on ATE's efforts to create advanced technological education is significant. Although some adjustments can improve the collaborative effort, clearly the objective of partnering educational deliverers and business/industry to produce a world-class workforce is a strong point of the ATE program. Based on analyzed data, the following specific conclusions seem reasonable concerning ATE centers'/projects' use of collaboration:

- The use of collaborative arrangements by *projects*, especially community college based *projects*, is widespread and a fundamental characteristic of the organizational entities involved in leadership roles.
- The ATE *project* was the initiator of the collaborative arrangements.
- The collaborative efforts included appropriate members for the intended outcomes.
- The greatest numbers of *project* collaborations were formed around business/industry, K-12 (primarily secondary) schools, and community and/or four-year colleges.
- Most collaborations with business and industry seemed to focus on workplace standards development and work-based educational experiences.

- Business and industry were not normally involved in verifying the validity of newly developed curriculum products.
- All reported collaborations had a focus on the *project's* goals, objectives, and desired outcomes. The degree that these were shared with partners varied, but in general, there was an adequate awareness of the ATE *project's* focus.
- The reported data did not indicate that a clearly defined mutual benefit to *project* collaborators had been routinely established. This was particularly true in collaborations with business and industry and with K-12 schools.
- The reported collaborations varied widely in duration and formalization. The range was from short, ad hoc relationships of a few days to long-term, sustained relationships based on formal agreements.
- Since those data gathered were from currently active projects, there is no way to judge the persistence of the collaborations after grant funding.

RECOMMENDATIONS

Based on these conclusions, the following recommendations are provided for NSF and ATE *projects*.

Recommendations for ATE

(1) *Project* proposals should clearly identify the expected collaborations to include types of membership, duration, purpose(s), and expected outcomes.
Since collaboration is a major element of successful ATE projects, it is important for ATE to understand with whom, why, and how those proposing a *project* intend to proceed with partnering. By specifying that proposals include a section describing intended collaborations, ATE can ensure that the *project* is approaching collaboration in an appropriate way.

(2) ATE should encourage statewide, regional, or national consortia as part of the collaboration model for large *project* awards.
To obtain the "biggest bang for the buck," ATE *projects* need to reach the maximum number of users possible for their products. Data gathered during site visits indicate that collaborative arrangements involving multiple educational deliverers have the greatest potential for product spread and adaptation.

(3) ATE should actively facilitate center-to-project collaboration and, where similar outcomes are identified, project-to-project collaboration.
Data indicate there is little collaboration between centers and projects with similar expected outcomes. Projects are significantly smaller and have limited

resources to disseminate their products, and the greater reach of centers can enhance wider dissemination. One possible approach is for ATE, when funding their activities, to provide *projects* with a listing of specific potential partners and task centers to nurture a collaborative arrangement with projects. Since centers' funding spans a significant period of time, periodically updating the list of potential project partners would be required.

(4) For *projects* involving development of technical curriculum products, business and industry involvement in verification of the product effectiveness should be required.
Projects are doing a creditable job in enlisting business/industry support for most technical education activities. The only issue noted in reported data is in the limited use of business/industry expertise in validating products or programs. ATE should require *projects* to include business/industry representatives in pilot/field-testing activities as part of the evaluation of *project* success in developing and implementing products associated with technical work-based knowledge and skill.

(5) ATE should provide guidance on best practices for sustaining collaborations beyond *project* funding and systematically gather data concerning collaboration sustainability.
Some collaborative arrangements engaged in by *projects* are dependent on funding (e.g. sustaining meaningful contact with other educational deliverers) and may not persist past ATE funding. Since technical programs are dependent on continual improvement to stay abreast of the needs of business/industry for skilled technicians, ATE should provide *projects* with guidance on how best to ensure that these collaborations continue after funding ceases. As part of the overall evaluation of ATE, data on the degree collaborative activities persist after funding should be gathered. This should be part of an overall evaluation strategy that looks at the post *project* impact of ATE.

Recommendations for ATE Centers and Projects

(1) Establish sustained relationships with business/industry throughout the *project's* funded life and beyond.
(2) Clearly identify the benefit of cooperation and support that accrues to collaborating partners.
(3) Formalize membership, objectives, procedures, and anticipated outcomes for collaborations.
(4) Facilitate appropriate levels of contact with collaborators, and seek meaningful input toward goals, and keep them informed of progress.

(5) Clearly identify the NSF-ATE role in the *project*.
(6) Involve businesses and industries in verifying the efficacy of technical curriculum products.
(7) Form collaborations that extend the reach of the *project*.
(8) If a collaborative relationship is not working fix it or disband it.

SUGGESTED APPROACH TO EVALUATION

Below are 14 questions for use in evaluating ATE-supported collaborations. The answers to the questions can provide a matrix of data on which to judge the collaboration's purpose, membership, persistence, and effectiveness in meeting goals of the ATE-supported *project*.[29] These questions can also serve as a framework for planning and designing collaborations that have a high probability of success.

(1) What is the purpose of the collaboration?
(2) Does the purpose clearly relate to the goals of the ATE *project*?
(3) Is the purpose of the collaboration defined, and are partners aware of the purpose?
(4) Is the membership of the collaboration appropriate for the purpose?
(5) Do the partners in the collaboration understand the ATE involvement?
(6) Does the collaboration provide mutual benefit to members?
(7) Is there a formal structure for the collaboration?
(8) Was collaboration initiated by the ATE *project*?
(9) If the collaboration predated ATE funding, is there evidence that ATE resources have strengthened the collaboration?
(10) Does the ATE *project* routinely communicate with and keep partners involved in activities related to the stated purpose?
(11) Is the collaboration a long-term or a limited relationship?
(12) How effective is the collaboration in achieving its stated purpose?
(13) Does the collaboration link the *project* with partners that provide an expanded network for integrating the *project's* products into workforce programs beyond the service area of the *project* organization?
(14) What is the probability that successful collaborative efforts will persist after ATE funding ceases?

NOTES

1. *Status Report 1: The Nature of the ATE Program*, p. 2, Table 1, Kalamazoo, Michigan: The Evaluation Center, Western Michigan University, May 2000.

2. *Advanced Technological Education, Program Announcement 97-29*, NSF Web site http://www.ehr.nsf.gov/EHR/DUE as of March 27, 2001, pp. 2–4.

3. Reflected in NSF program application guidance: *ATE Program Announcement*, 1998, *ATE Program Solicitation* 2000 & 2001. Other references to the 1997 guidance contained in citation 4 above are also reflected in these later documents.

4. Literature was selected based on Internet and university library searches. Selected items focused on collaboration and partnerships in an educational setting.

5. The Evaluation Center at Western Michigan University has been designated by NSF to evaluate the degree of goal achievement, impact in reaching intended individuals and groups, effectiveness in reaching constituents, and recommendation on possible significant improvement.

6. Upon review of this paper, NSF suggested a hierarchy of contact, collaboration, and partnership. Contact is making a presentation or getting funding. A collaboration is working on a joint endeavor for a short time with out many agreements. A partnership is a longer-term relationship with definite expectations on both sides.

7. James R. Mahoney (Ed.), *Improving Science, Mathematics, Engineering, and Technology Instruction, Strategies for the Community College*, 32-4, Washington DC: Community College Press, 1996.

8. As a matter of convenience the term *community college* will be used in this paper when referring to associate-degree-granting colleges.

9. Charles R. Monroe, *Profile of the Community College*. San Francisco: Jossey-Bass Publishers 1973, pp. 78–87.

10. "Interdependence Through Partnerships: Transforming Education," *Improving Science, Mathematics, Engineering, and Technology Instruction, Strategies for the Community College, 64*, Washington DC: Community College Press, 1996.

11. *Ibid.* p. 65.

12. "Advancing Technological Education," *Synergy*, p. 4, Arlington, VA, National Science Foundation March 1999.

13. "1997 Awards and Activities," *Advanced Technological Education* p. 1, Arlington, VA, National Science Foundation 1997.

14. *Advanced Technological Education, Program Announcement 97-29*, ATE, pp. 2–3.

15. James R. Mahoney, Lynn Barnett (Eds), *The Learning Edge, Advance Technology Education Programs at Community Colleges*, 7–14, Washington DC: Community College Press, 2000.

16. ATE Site Visit Report Outline, *ATE Drivers*, The Evaluation Center, Western Michigan University. Kalamazoo, Michigan, Distributed September 2000.

17. Data highlighted in this paper are from the 2000 survey report. If there are significant difference between the 2000 and 2001 data, these differences are referenced either in the body of the text or in an appropriate footnote.

18. Site visit reports are comprehensive documents created by members of The Evaluation Center's "visiting teams" that did on-site reviews of ATE projects at 13 locations. These reports have limited distribution to preserve the anonymity of the sites visited. Therefore, no citations will be provided when referring to information contained in these reports.

19. *Status Report 2: Findings From a Survey of ATE Projects and Centers*, p. iv.

20. *Ibid.* p. iv.

21. *Survey 2001: The Status of ATE Projects and Centers*, p. 48.

22. 2001 data. 2000 data showed a lower participation.

23. *Ibid.* 20.

24. *Ibid.* 21-22.

25. *Profile of the Community College.* pp. 29–32.

26. The models presented are a synthesis of data reported by visited sites. Although several sites primarily used one of the presented models, other sites exhibited characteristics of portions of several models.

27. Collaboration is a process to bring about an outcome. In this case, the overall outcome is dissemination of the benefits of improved technical education. Subobjectives of the collaborations described in this paper often include materials development, resource support, recruitment and placement, etc. In other papers describing the ATE program, the formation of collaborations may be addressed as means to a specific end. The models presented are offered as an overall approach to expanding the influence of ATE.

28. *Advanced Technological Education, Program Announcement* 97-29, p. 2.

29. The profile created from answers to these questions should be compared with the definition and collaboration elements discussed above to determine the degree to which a *project* partnership is meeting the intended goal of successful collaboration.

DISSEMINATION

Thomas Owens

INTRODUCTION

While the history of innovation in American education is rich with new ideas and projects that have benefited local communities, there has not been a continued systematic effort to disseminate most of these ideas and practices to educators in other parts of the country. There is even less success in helping other educators and communities to adapt these ideas and practices to meet their local needs. Because federal agencies such as the National Science Foundation (NSF) have invested billions of dollars in programs and projects to help improve education, there is a need to give greater attention to disseminating the policies, practices, and materials that have already been developed and to assisting others in adapting them to improve education in their communities.

LITERATURE REVIEW HIGHLIGHTS

Definitions and Conceptualizations of Dissemination

This brief literature review focuses on ideas considered especially relevant to the NSF ATE staff and those in the ATE *projects* (i.e. centers and projects). Emphasis here is given to factors found to facilitate effective dissemination. The review does not include the history of dissemination nor the philosophical underpinnings for dissemination. These have been covered well elsewhere (see Hutchinson & Huberman, 1993; Louis & Jones, 2001).

Evaluating the Upgrading of Technical Courses at Two-Year Colleges:
NSF'S Advanced Technological Education Program
Advances in Program Evaluation, Volume 9, 35–56
Published by Elsevier Ltd.
ISSN: 1474-7863/doi:10.1016/S1474-7863(04)09003-9

The term *dissemination* has different meanings to different people. According to Hutchinson and Huberman (1993), "Its most common definition is the transfer of knowledge within and across settings, with the expectation that the knowledge will be 'used' conceptually (as learning, enlightenment, or the acquisition of new perspectives or attitudes) or instrumentally (in the form of modified or new practices)" (p. 2).

Another definition is that "Dissemination consists of purposive, goal-oriented communication of information or knowledge that is specific and potentially useable, from one social system to another" (Louis & van Velzen, 1988, p. 262).

Several theories or frameworks were reviewed that add insight into dissemination by ATE *projects*. One of these is the work of the Dissemination Analysis Group done at a conference of dissemination professionals (Klein, 1992). This group identified four functions of dissemination and appropriate strategies for achieving each.

(1) Spread – the one-way broadcasting of information, in order to increase awareness.
(2) Choice – the provision of information on options intended to help users compare alternative resources.
(3) Exchange – interaction of information, materials or perspectives.
(4) Implementation – technical assistance, training or other forms of support to change attitudes or behaviors and to institutionalize changes over time.

Hutchinson and Huberman (1993) show how different activities are appropriate for each function of dissemination. For example, publications, presentations, and telecommunications may be appropriate for spread. Choice is facilitated when a person responds to a client request or query that can lead to the client understanding the advantages and disadvantages of an option. Exchange is helped by individual meetings, workshops, and seminars. Implementation is fostered through direct assistance, training, and sustained support for change. The ATE *projects* may find that all four functions are appropriate for their use depending on the specific needs of individual clients at the time.

Factors Considered Facilitating or Hindering to Effective Dissemination

A review of the literature on dissemination reveals five areas in which there are practices that either facilitate or hinder effective dissemination: (1) the information users; (2) quality of the information; (3) adaptability of the information; (4) diverse modes of communicating the information; and (5) support for utilization. Table 1 lists practices that have been identified in the literature as having facilitated or hindered dissemination.

Table 1. Factors Found to Facilitate or Hinder Effective Dissemination.

Areas	Facilitators	Barriers
1. Information users	Understand characteristics of the users Use preferred language style of the users	Poorly targeted groups Inadequate information about the users
2. Information	Timely Comprehensive Accessible Validated materials/practices	Insufficient evaluation of the materials to be disseminated Low quality materials/practices
3. Adaptability	Users can easily adapt materials to their needs Materials are seen as meeting the users' needs and concerns	Lack of attention to the need for users to want to adapt materials/ practices to their local settings
4. Diverse modes	Includes electronic, print, and person-to-person communications Interactive	Only one mode used Reliance on one-way communication
5. Support for utilization	Ongoing interactions with users Dissemination is integrated with other R&D functions Uses networks for dissemination Has training and technical assistance to match user needs	Limited local development and training Inadequate structure for between-group sharing

As part of its theoretical framework, the National Center for the Dissemination of Disability Research (NCDDR) has identified eight premises based upon the findings of research and experience. Their premises support the common factors listed above. In addition, they stress that:

- Dissemination and distribution are not the same.
- Effective dissemination is not an "end activity" that occurs after research is completed.
- Recipients of government-funded research funds have a responsibility to effectively disseminate their results (NCDDR, 1997).

A comprehensive review of the literature in knowledge dissemination and use in science and mathematics education was done by Hutchinson and Huberman (1993) for the Directorate of Education and Human Resources, Division of Research, Evaluation and Dissemination within NSF. Their experience and review of earlier research in various disciplines call attention to seven factors that facilitate dissemination:

(1) Accessibility, availability, and adaptability – easy access to information that can be locally adapted.
(2) Relevance and compatibility – the information being disseminated is seen by practitioners as fitting their world and concerns.
(3) Quality – materials have been evaluated or validated.
(4) Redundancy of the messages – repeated messages over time and through diverse dissemination modes.
(5) Linkage among users – interpersonal interactions among users.
(6) Engagement – opportunities for users to engage with the new materials or ideas.
(7) Sustained interactivity – frequent contact between information users and providers.

Westbrook and Boethel (1997) found that successful dissemination systems have the characteristics identified in Table 1 in common. In addition, they indicate that successful dissemination systems:

- Include both proactive and reactive dissemination channels – that is, they include information that users have identified as important, and they include information that users may not know to request, but that they are likely to need. Clear channels are established for users to make their needs and priorities known to the disseminating agency.
- Recognize and provide for the "natural flow" of the four levels of dissemination that have been identified as leading to utilization: spread, exchange, choice, and implementation.
- Draw upon existing resources, relationships, and networks to the maximum extent possible while building new resources as needed by users.

In the 1960s, Havelock and other members of the Institute for Social Research at the University of Michigan proposed a RDDE cycle consisting of Research, Development of prototypes, Diffusion of the amended prototypes, and Evaluation of the product (Havelock, 1969). This model was important in the creation of the educational laboratories and centers created by the U.S. Department of Education. During the 1970s, other researchers identified problems with the RDDE model. Later, others disagreed with the flow of knowledge as a one-way process that did not take into account the motivations, contexts, and realities of the intended users.

Hutchinson and Huberman (1993) described the shift from the one-way flow models to a constructivist perspective in which "the user acts upon information by relating it to existing knowledge, imposing meaning and organization on experience and, in many cases, monitoring understanding throughout the process. This casts the user as an active problem-solver" (p. 2).

The most frequently cited reason for the gaps between research and its use center on the lack of communication and cooperation between researchers and their intended audiences (Leung, 1992).

Smink reviewed a series of federal dissemination studies. He found that there were many operational problems with dissemination. Each of these was already included in the common barriers in Table 1. He also found weak incentives for use among practitioners (Smink, 1985).

Dissemination Guidelines and Tools

Federal guidelines related to dissemination also contain insight into factors that contribute to successful dissemination. For example, The Fund for the Improvement of Postsecondary Education (FIPSE) emphasizes that dissemination must be adapted to the particular circumstances of the new environment and that it is important to establish that those wishing to adapt the reforms are ready to take action. It also stresses the importance of providing training and ongoing implementation assistance, having a period long enough to permit extensive coaching, and being proactive in communicating regularly with adapting sites to sustain and support progress (FIPSE, 2000).

The NSF Dissemination Partnership Program involves an institution-to institution mentoring approach. Its guidelines state that "Disseminators have learned that their efforts yield the strongest and most lasting results when the project includes some of the following:

- A good product with proven or promising results.
- A match between the experience and knowledge of the grantee and the needs of the partner institutions and/or agencies.
- A mutual understanding that the promising practice or component will be adapted to at the particular circumstances of the partner institutions and/or agencies.
- Substantial involvement of the partner institutions and/or agencies in the developing of the application.
- A readiness on behalf of the partner institutions and/or agencies to take action
- An action plan which includes on-site technical assistance.
- Systematic contact and communications between the grantee and the partner institutions and/or agencies, including face to face contact.
- Clear roles and responsibilities between the project and the partner institutions and/or agencies.
- Well defined objectives for the project.

- A strong evaluation plan that will document the effectiveness of the practice (or program component) at the adapting sites (U.S. Department of Education, Office of Postsecondary Education, 2000).

In addition to the factors associated with effective dissemination identified above, some strategies and tools developed through federal funds have facilitated effective dissemination. One of these is the *Dissemination Self-Inventory*. This self-inventory was developed to assist National Institute on Disability and Rehabilitation Research (NIDRR)-funded project staff in reviewing their dissemination practices and is available on line at www.ncddr.org/du/products/tsi/index.html. It was designed to help guide the planning and implementation of dissemination by busy staff with limited time and resources (NIDRR, 2000).

The self-inventory contains rating scales for items grouped into five categories as shown below with a sample item from each:

User group – user group(s) or potential users of the information or product to be disseminated. Example: Does your research design clearly define the intended groups of "users" or beneficiaries of your project's results?

Information source – your project/organization as an information source, that is, the agency, organization, or individual responsible for creating new knowledge or products, and/or for conducting dissemination activities. Example: Are your project staff regarded by users as highly knowledgeable resources in the project's topical area?

Content of message – message content that is disseminated, such as, the new knowledge or product itself, and any supporting information or materials. Example: Does your project information contain examples or demonstrations of how to use, and the implications of use, of the information?

Medium of the message – the ways in which the knowledge or product is described, "packaged," and transmitted. Example: Does your project make information available in any alternate format requested by individual users?

Contextual consideration for implementation – context for use of the message, that is, the environmental, personal, and other supports needed to use the information or product. Example: Does your project develop a written plan with objectives as a guide in delivering technical assistance to user groups?

The *Dissemination Self-Inventory* is based on the research literature on dissemination, knowledge utilization, and the change process. In addition to scoring directions, the instrument contains a useful set of references organized around the five categories. These items could easily be adapted for ATE *projects*.

FINDINGS FROM THE WMU EVALUATION PROJECT SURVEYS AND SITE VISITS

The WMU evaluation project 2000 and 2001 surveys (Gullickson et al., 2000, 2001) addressed dissemination through one question in the PI overview section, which asked for product dissemination methods. There was also reference to one aspect of it under materials development.

In each year of the survey, more than 1,000 of the materials developed were reported in use at least locally. If one presumed that all developed materials were used at least on a local basis, then in each year at least 35% of this total were used at sites other than the *projects*, and 11% were commercially published. It should be noted that some of these materials were modules vs. course development or course adaptation. Thus, *projects* may have reported modules both separately and as part of course development or adaptation materials. Despite the potential this raises for over-reporting, this author suspects that the figures overall are an understatement of use of materials since many *projects* may be unaware of some sites that may use their materials.

The site visits to 13 selected ATE *projects* conducted by the WMU evaluation project did not systematically address the issue of dissemination as a specific topic, but it was imbedded within materials development or grouped with sustainability and transportability. A review of the 13 reports indicated that dissemination was mentioned as a *project* objective at 5 of the 13 sites. It must be noted that it does not mean that dissemination was not important just because it was not mentioned during the site visit.

The most frequent mention of dissemination was in regard to presentations made by *project* staff at professional conferences. Several sites combined the Internet with the use of CDs as a platform for their curriculum. One site mentioned use of networking groups with educators and business members to communicate the importance of their curriculum emphasis. Other dissemination activities mentioned included a pending article in the *Journal of SMET Education*, development of PowerPoint presentations, arrangements with a commercial publisher to disseminate modules, the dissemination of emerging technology trends, the adoption of one curriculum by 15 colleges, and a clearinghouse and network for continuing collaboration of partners regarding their *project*'s area of focus.

COMPREHENSIVE DISSEMINATION EXAMPLES

While it is useful to review prior dissemination studies and findings from national surveys and site visits, it may also be useful to have a few examples of comprehensive dissemination approaches. Two examples are presented here that suggest individual strategies or sets of strategies that could be used by ATE projects (especially the larger projects) or ATE centers. The first comes from the new National Dissemination Center for Career and Technical Education (NDCCTE), and the second is an example of comprehensive dissemination activities at one of the ATE centers.

NDCCTE. Perhaps the largest dissemination contract in vocational technical education in the U.S. was awarded in 1999 to NDCCTE as a five-year contract for implementation between 2000 and 2004. The national dissemination is being implemented by a consortium of primary partners (The Ohio State University – Prime, University of Minnesota, University of Illinois, The Pennsylvania State University, and Oregon State University with assistance from Johns Hopkins University and the Academy for Educational Development) (McKinney et al., 1999).

The dissemination strategies being used by NDCCTE include those listed below:

Web site available for all users.
All documents and publications available in electronic format on the Web site.
Print documents available for those with limited or no access to the Internet.
Information that is easy to use, featuring short summaries, well-designed graphs and charts.
Information made available through a variety of channels, including person-to-person communication through the Web, teleconferencing, and interactive dialogue between current users of an idea and potential users of an idea.
Information presented in varying depth and length, depending on the nature of the target audience.
Information presented using videos and electronic media to reach those more favorably inclined to receive information through a visual medium.
Provision for access to information at a time when it is relevant to the user via the Web site and the Question-and-Answer Service.
Information made available through multiple channels, including print, audiovisual, electronic, and person-to-person.
Presentations at significant conferences and institutes (McKinney et al., 1999, p. 214).

Some principles used in the planned external evaluation of NDCCTE also have relevance for the evaluation of ATE centers and larger projects. A logic model guides the assessment of dissemination impact. The logic model is organized around six questions each supported by indicators/criteria, standards, and procedures.

(1) Is the center establishing the preconditions for impact?
(2) Is the center reaching its intended audiences?
(3) Is the center maintaining national visibility?
(4) How do the clients and the field perceive the outputs of the center?
(5) Have the knowledge, skills, and/or behaviors of clients changed as a result of center activities?
(6) Have center products and services had an effect on policies and/or practice in the field? (Altschuld, 1999)

Procedures planned in the NDCCTE evaluation include analysis of records and database, an annual survey of a sample of center clients, requests for center products, evaluation of specific products and services, case studies and tracer studies, staff surveys, and a survey of specially constituted expert panels. The ATE centers and larger projects may want to adapt the above six questions and include them at various times during their operations.

In applying the 14 factors considered facilitating for effective dissemination as shown in Table 1, the NDCCTE model addresses 12 factors. The 2 that are not clearly addressed are the check on validated materials/practices and the available training and technical assistance to match user needs.

NWCET. At the National Workforce Center for Emerging Technologies, the new name for the NorthWest Center for Emerging Technologies (an ATE center for information technology at Bellevue Community College), dissemination is an essential ingredient in providing national leadership.

What are a few examples of some dissemination strategies they are using? Illustrations are drawn from its latest NSF report (NWCET, 2001). First, the center listened closely to its customers before acting. This included partnering with the American Electronics Association to do industry-expert reviews nationwide that served as the basis for validating and updating the *Millennium Edition of the NWCET IT Skill Standards*. They also worked with the external evaluator to conduct studies of educators and businesspeople who used the earlier version of the Skills Standards to determine how they used the standards and ways they felt the document could be improved.

To help address the question of what it would take to attract middle and high school youth, especially women and minorities to the IT field, NWCET contracted with a marketing firm to do focus groups in various parts of the country. These consisted of panels of students, parents, and educators and were used to establish the research foundation for producing the Cyber Careers for a Net Generation video and classroom materials.

Recognizing that there are many developers of IT Skill Standards-based products and services, NWCET established a compliance review process that helps ensure that developers across the country produce valid, quality courseware, assessments, and related products.

In addition to regular presentations about the center and its best practices to local, state, and national conferences of educators and industry leaders, it also organized a Partners Summit to bring together top executives and leaders in education, business, and technology to discuss trends impacting IT education and workforce issues and to learn about NWCET initiatives. Recognizing that the largest bottleneck in the IT workforce shortage is the supply of qualified IT instructors, the center launched the Educator-to-Educator Institutes across the U.S. with instructors who are certified by NWCET. And, finally, center staff are willing to reach out to influence and assist other groups. For example, the center's director testified before the U.S. House of Representatives Committee on Science on behalf of NSF funding requests. The associate director advises the Information Technology Association of America, serves on the National Academy of Sciences National Research Council, and is coauthor of their report to the U.S. Congress on IT workforce shortage issues.

In applying the 14 factors considered as facilitating effective dissemination as shown in Table 1, the NWCET model addresses all 14 factors. How have they done it? Table 2 shows examples of the 14 elements.

Table 2. NWCET Examples of Factors Facilitating Effective Dissemination.

Areas	Facilitators	Examples
1. Information users	Understand characteristics of the users	NWCET identified four specific audiences: educators, students, employers, and government staff. Its Web page, for example, is geared for quick access to information of special interest to each of these audiences.
	Use preferred language style of the users	The Cybercareers for a Net Generation, a video and support materials to interest young people in Information Technology (IT) careers, was based on a prior careful study of young people's attitudes toward and misinformation regarding IT. It then videotaped young people of color discussing issues related to IT.
2. Information	Timely	The IT skill standards are updated every several years, and information on the Web site is updated widely.
	Comprehensive	Information and materials have been developed that address all areas of IT workforce training including IT standards, curriculum, training of educators, orientation of young people to IT, and even addressing the need to recruit and retrain more IT educators.
	Accessible	Many of the materials are immediately available on the Web site as well as in print.

Table 2. *(Continued)*

Areas	Facilitators	Examples
	Validated materials/practices	Materials are based on IT standards that were recently validated nationally.
3. Adaptability	Users can easily adapt materials to their needs	The NWCET staff adapted their materials for special groups such as the Job Corps and have provided a model and technical assistance to others in adapting materials to their state or local needs.
	Materials are seen as meeting the users' needs and concerns	Evaluation surveys of educators and industry people who have used the IT skill standards have indicated how the standards are meeting their needs.
4. Diverse modes	Includes electronic, print, and person-to-person communications	NWCET used its Web site, printed materials, training sessions, testimony to congressional committees, and many conference presentations.
	Interactive	The center staff use continuous contact with their clients to obtain feedback regarding their products, services, and training.
5. Support for utilization	Ongoing interactions with users	The quarterly meetings with the National Advisory Board have been examples of interactive exchanges among the staff and advisory board members representing education, business, and government.
	Dissemination is integrated with other R&D functions	Dissemination is closely coordinated with staff training, curriculum development, research, and evaluation.
	Uses networks for dissemination	The center uses its contacts with other community colleges to help expand curriculum development and training. A contract with AACC is used to disseminate the IT skill standards. Educators and business leaders on the NAB have been active in disseminating information about the center to their institutions and colleagues.
	Has training and technical assistance to match user needs	Under separate contracts, NWCET trained staff from all the other community colleges in Washington. Through support from the Microsoft Corp, NWCET is setting up training centers throughout the U.S. to help train IT educators.

FINDINGS FROM AN ATE CENTER DIRECTOR SURVEY AND WEB SITE ANALYSIS

ATE Center Director Survey

To supplement information from the literature review, annual reports, and site visits to ATE *projects*, the author conducted an e-mail survey of the ATE center directors about their dissemination practices. A draft instrument was sent to an ATE center director and associate director for their feedback. Following the feedback, in May 2001, the survey was sent as an e-mail attachment to the directors of the 11 centers. Responses were received from 9 of the 11 directors (1 director had retired and another center was not currently functioning). A summary analysis of the responses was prepared by this author and is shown below.

The ATE center director survey consisted of eight questions listed below. Responses to this survey are shown here, and highlights are integrated into other sections of this paper. NSF could use these eight questions as part of their site visits to ATE centers or large projects. They may not be relevant to small projects.

(1) What have been your major strategies for disseminating policies, strategies, and materials (such as curricula) developed by your ATE center?
(2) Which dissemination strategies do you feel have worked especially well? Why?
(3) Have you used any strategic partnerships to help with dissemination? If yes, which groups? How effective do you feel these partnerships have been? Why?
(4) What are the primary types of evidence you have to support the impact of your center's dissemination efforts?
(5) Which dissemination strategies do you feel have not worked too well? Why?
(6) Based on your experiences with dissemination, what new strategies might be tried by your ATE center in the future?
(7) What changes or additions would you support for the NSF guidelines regarding dissemination?
(8) What ways could NSF or the ATE centers use for sharing "best practices" in effective dissemination?

1. What have been your major strategies for disseminating policies, strategies, and materials (such as curricula) developed by your ATE center?

ATE center directors most frequently mentioned using their ATE Web sites, exhibits, presentations at national conferences, word of mouth, and partnerships with other organizations. Other strategies used included providing Congressional testimony, use of state or regional partners, partner meetings, serving on other ATE national visiting committees, serving on education or industry advisory boards, brochures, career days/fairs, student competitions, preparation of best practices, faculty development workshops, special topic publications such as on student retention, using informed college students and employers as spokespersons, online or printed newsletters, electronic mailings lists, journal articles, and production of videotapes especially for students.

2. Which dissemination strategies do you feel have worked especially well? Why?

The most successful strategies were targeted dissemination efforts including faculty development workshops because there is more opportunity for one-on-one attention and interest from recipients. This allows center staff more opportunities to work with faculty to ensure effective use of the new curricula or other strategies. Some specific curriculum adopter workshops were co-hosted by partner schools with live interactive practice.

Use of Web sites was found to be less effective as a stand-alone strategy, but helpful, when employed in actual contact with users. The Web sites were also found to be particularly useful when organized for easy access by targeted audiences such as educators, students, business, and government. Cross-referencing of ATE center Web site information from other education and industry leading organizations also added credibility.

Using knowledgeable industry leaders and current or recent community college students who had engaged in the technology programs was also found persuasive with many audiences.

Although print and electronic newsletters may not lead to specific documented changes, they were found effective in reaching larger audiences for awareness purposes.

One center recently experienced success when curricula and materials were delivered via a hybrid web and CD-ROM system that allowed updating through the Web but provided high bandwidth items via the CD.

State and regional partnerships were also noted as effective in helping to share information and tailoring it to local needs.

Several directors mentioned the effectiveness of mentoring projects where they work with other community/technical colleges in developing and/or adapting instructional materials and models for use in their targeted area of technology.

3. Have you used any strategic partnerships to help with dissemination? If yes, which groups? How effective do you feel these partnerships have been? Why?

Use of business and education partners to assist with dissemination was common among ATE centers. Regional centers and state departments or organizations were especially effective in connecting to high schools, colleges, universities, and industries in the local areas. State Departments of Education and Commerce were mentioned as partners. Technology alliances and industry associations were also mentioned, as were commercial publishers who assist with curriculum dissemination.

Involvement with professional societies and groups like the League of Innovation were also found helpful, as were contacts initiated by the NSF staff.

4. What are the primary types of evidence you have to support the impact of your center's dissemination efforts?

A variety of evidence was cited to support the impact of dissemination including (1) the types and numbers of requests from the field for information, curriculum, materials, training, and proposal development; (2) number of students enrolled in ATE classes, graduating, receiving ATE scholarships, being hired as technicians, and promoted in their technical fields; (3) number of colleges offering the new ATE center courses, using curriculum and marketing/recruiting materials, and expressing interest in adapting ATE approaches; (4) follow-up evaluations of faculty, students, and industries using products or services of the centers; (5) regular Web site usage reports; and (6) reports from partners.

5. Which dissemination strategies do you feel have not worked too well? Why?

In the words of one center director, "Just placing materials on a Web site is not working. Other strategies must be use to create a 'need to know' to grow a Web site audience." Another said, "Simply telling people about our materials and delivery system is inadequate. They need to see the system demonstrated to appreciate the multimedia features and ease of use."

Other barriers mentioned included faculty concern regarding intellectual property ownership of instructional material they develop that get placed on the Internet.

Mass mail-outs were found to be of limited usefulness by some center directors.

It was also mentioned that most colleges are reluctant to use newer student recruiting strategies (e.g. career fairs, news articles, and newspaper advertising) and to hire contract help to spearhead recruiting efforts.

6. Based on your experiences with dissemination, what new strategies might be tried by your ATE center in the future?

Responses mentioned included development of general interest videos for students, focusing more on ATE outcomes, a cost/benefit analysis, more targeted marketing, more on-line dissemination, use of Web-based surveys of faculty and industry, identifying and presenting at new conferences such as those that attract high school teachers, grants to fund other community/technical colleges to develop or adapt new technology curricula, setting up a clearinghouse for other information and curricula in the respective technology area, use of streaming audio and video CDs, and experimenting with PDF files and user passwords to allow follow-up and prevent pirating of materials placed on the Internet. The overlap of these techniques with ones that have been unsuccessful in other settings (e.g. use of the Web) points out the need for careful targeting of the dissemination strategy.

7. What changes or additions would you support for the NSF guidelines regarding dissemination?

Although the suggestions might be more appropriate for a guidebook than the NSF Program Announcement, suggestions included NSF identifying some effective dissemination strategies, suggesting what to avoid, stressing that integrated strategies must be considered that include professional development, utilizing ATE centers as hubs for disseminating ATE project materials and findings, and establishing a NSF distribution and marketing center for ATE center and project products so proposal writers could borrow from what is working.

8. What ways could NSF or the ATE centers use for sharing "best practices" in effective dissemination?

Several directors suggested cross-training PIs and some professional staff of ATE centers and large projects to speak about not only their own materials but also other centers' outstanding materials and strategies. Other suggestions included using centers as clearinghouses for information on projects addressing specific topics where best practices are being developed, better use of technology to share "best practices," more sophisticated cataloging of work being done with search engines that would take the use of the type of data provided to the current FastLane to a new and more helpful level, sharing best practices at the annual NSF and PI conference and have them posted on a Web site bulletin board maintained by the centers, coupling best practices with professional development needed to help others learn how they can do it, and holding a facilitated discussion of NSF staff and center PIs on specific dissemination outcomes expected.

ATE centers Web Site Analysis

In addition to the literature review, data from the WMU evaluation project's 2000 and 2001 surveys, and survey of ATE center directors, this author reviewed each of the available 10 ATE center Web sites to identify types of information disseminated there. There was no attempt to judge the technical quality of these sites or their impact but merely to describe the types of information they were disseminating. On the home page, information was available specifically for students (at 5 sites), educators (at 9 sites), business/industry (at 4 sites), and for government (at 2 sites). Table 3 shows the number of ATE centers displaying various types of information. As shown in Table 3, all 10 ATE centers' Web sites contained information about their center, half included job openings in their technical field, and one included an electronic newsletter.

Table 3. Types of Information Displayed on ATE Center Web Pages.

Type of Information	Number of ATE Centers
Information about the center	10
Coming events/calendar	7
Curriculum	6
Job listings	5
Current news in the technical field	5
References/publications	5
Partner site information	4
Clearinghouse searches	3
Newsletters	1

A NEW PARADIGM FOR DISSEMINATION

The prior review of the literature and survey of ATE center directors makes clear that some of the past ideas of dissemination need to be updated to accommodate future needs of educators, industry, and the public. No longer can dissemination be viewed as a one-shot activity, flowing only from the centers to users, or separated from other components of successful ATE *projects*. It must be ongoing, planned, and implemented in continuous dialog with the information/product/service users and comprehensive enough to address the variety of interrelated needs of the customer. Professional development of clients in the form of training and technical assistance is an essential component of the newer view of dissemination. It must also help anticipate the future needs of users in solving their problems or creating new opportunities for them. These elements have been supported by the literature review as shown in Table 1.

Two concepts from business and e-commerce have applicability as we contemplate a new paradigm for dissemination at ATE centers – integrated solutions provider and customer relationship management. Each is described here, followed by a new definition of dissemination that incorporates implications from these two business concepts.

The Director and Associate Director of the National Workforce Center for Emerging Technologies (NWCET) (called the NorthWest Center for Emerging Technologies before June 2001) shared a concept emerging from e-commerce that has relevance – an integrated solutions provider (ISP).

As Dr. Peter Saflund, Associate Director of NWCET has told me recently, in regard to an integrated solutions provider:

> It's sort of like Wal Mart. The more you can put under one roof, the more likely your products and services are to synergize. As an Information Technology example: if I design Web pages, but

> also procure graphics, host the site for my clients, offer shopping cart and transaction processing, offer to help my clients design and implement web marketing services, and possibly integrate suppliers or complimentary goods I'm an integrated solutions provider.
>
> Conceptually, it's the difference between merely 'selling cell phones' and offering services to 'keep people connected.' The latter concept includes family discounts, flexible calling plans, voice and text messaging, and maybe other services to help the client get the most out of his/her phone.
>
> So, listening to how your clients are using your products and services – figuring out what problems they have that your products and services are solving for them, and being aggressive about how to help them use your products or services more effectively and productively – that's the essence of solutions-oriented marketing.
>
> In our case, we know almost nobody uses the Information Technology Skill Standards just as they are – there is almost always some interpretation, adaptation, application to existing curricula, needed. Rather than simply be reactive to these needs, we will do more to anticipate them and offer the solution package at the time of 'sale' (Personal e-mail communication, May 7, 2001).

For NWCET, the needs of the field have included a better understanding of the Information Technology (IT) workforce shortage and its causes, identification of occupational areas in the IT field (including those requiring less than a baccalaureate degree), nationally agreed-upon IT skill standards that are acceptable to industry and education, new IT curricula at the high school and college levels, adaptation of the IT curricula for special populations such as the Job Corps members, ways to help educators adapt existing IT curricula and certify IT programs as being IT skills standards compliant, processes for authentically assessing IT competencies in students and employees, ways to articulate community college and university programs for some areas of IT, strategies to interest middle school and high school students in IT, ways to attract more women and minorities into the IT workforce, ways to attract and retain IT faculty and upgrade their IT skills, strategies to effectively work with a national advisory board and other partners, and mechanisms for effectively sharing what was being learned and developed with educators and industry leaders across the U.S. and in other countries. NWCET has addressed each of these areas through its basic NSF center grant and augmented this with additional grants from NSF, Microsoft, and other sources including the Department of Labor.

Given the above set of needs, how does NWCET plan its dissemination? This ATE center announced that it "strives to be the premier National IT Education Solutions Provider and E-Portal Dissemination Point." The E-Portal to IT Education and Careers will be a new NSF-ATE dissemination focal point grant that will become a primary source for IT information, trends, programs, and best practices. It will be interactive, adaptive, and service oriented. NWCET will use the League for Learning Network and E-Portal to IT as marketing vehicles to extend the NWCET image and brand awareness.

Solutions to the problems listed above were not undertaken in isolation but in an integrated way that included work with numerous education and business partners with expertise in one or more of the above areas. In this sense, dissemination is an integral part of the entire research and development agenda of NWCET.

In industry, integrated solutions providers are becoming more common. For example, the United States Automobile Association (USAA) incorporated some of the above principles into what industry is now calling *customer relationship management* (CRM) and is using sophisticated software techniques such as data mining to extract a large volume of information about their clients. Kathleen Khirallah, a senior consultant with Tower Group, defines CRM as "a sales and service business strategy where the organization wraps itself around the customer, so that whenever there is interaction, the message exchanged is appropriate for that customer. That means knowing all about that customer and what the profitability of that customer is going to be" (Curley, 1999).

USAA, one of the world's largest insurance companies, has a single technology company within the company to provide cross-cutting solutions to its 16 business units. The firm is moving forward with plans to create huge data warehouses where customer information can be mined for service and sales opportunities (Curley, 1999).

The implications of CRM as a tool for educational institutions such as the ATE centers are clear. The centers can no longer get by with broad scattering of information. They must know well their clients and their needs and find ways to anticipate and meet these needs. This will necessitate keeping accurate records on ATE center customers and sharing new information with them as new products and services are introduced that might interest them.

Combining ideas from the above two concepts with the dissemination roles that ATE centers have been playing can lead to a newer definition of dissemination. In this new paradigm, dissemination is "the process of knowing your clients and systematically providing them, either directly or in partnership with other organizations, with knowledge, strategies, products and support that can enable them to better solve their problems and enhance their delivery of effective technical education."

Borrowing business models to apply to public education has its benefits as well as dangers. On the plus side, it can more closely integrate dissemination with other R&D functions, encourage customer-driven planning, be interactive, and support the sustainability of ATE centers. Some dangers to be aware of include making educational decisions based primarily on profitability, creating the image of centers as businesses, disseminating only the products of the center and not those generated elsewhere, invasion of privacy in order to create profitable databases of customers, and selling customer databases.

Thomas Bailey, in reviewing a draft of this paper suggested that we "examine the dissemination activities of ATE *projects* from two perspectives: public good and private good. Policies, practices and materials on STEM (formerly SMET) education disseminated by ATE *projects* are public good since these efforts raise the STEM competency of the community of the nation's people. For some *projects*, however, materials disseminated are often private goods as well as public. Dissemination in this case could be seen as more akin to marketing, generating net revenue, and enabling recipients to be self-sufficient in their ATE operations." It is possible that both public and private "goods" would be addressed.

RECOMMENDATIONS

Strengthening the ATE Proposal Requirements for Dissemination

At the present time, the NSF guidelines on dissemination are rather vague and simply require a dissemination plan. NSF specifies that dissemination needs to be more than maintaining a Web site. The NSF Program Announcement for National Centers of Excellence mentions the expectation that centers "disseminate their products through commercial publishers, journals, conferences, workshops, electronic networks and other means"; but suggestions are not given regarding promising practices to develop and maintain an effective ongoing dissemination presence. An adaptation of the dissemination self-inventory by the National Institute on Disability and Rehabilitation Research should be considered as a tool to guide the thinking of NSF proposal writers.

NSF should also consider the advice of one ATE center director who suggested that "Encouraging the integration of dissemination with mentoring, faculty development, and other project activities, rather than treating it as a stand-alone activity, would strengthen dissemination efforts and encourage the spread of best practices and useful curriculum developments." The idea of centers being "integrated solutions providers," as described in this paper, might guide the thinking of new and existing ATE centers.

If dissemination is to be perceived as an important part of future *projects*, there may also be a NSF requirement that, in addition to providing simple statistics such as the number of faculty and students served, centers and major projects should report impact data on at least one major product or service they provide.

NSF may also want to establish a best practices in dissemination section on its Web site and request that grant writers refer to these practices and consider using or adapting some of them in their own proposals.

While it may be true that all knowledge, materials, and practices developed under NSF funding should be effectively communicated to other potential users, limited resources dictate that there be various levels of dissemination expected depending on whether those receiving funds are small projects (under $100,000), large projects, or multiyear ATE centers. For small projects, a minimal expectation may be that staff at the local institution share the findings, materials, or practices with others at their institution and that in their reports to NSF they provide sufficient information that will enable NSF to disseminate such information to others. Larger projects might also be expected to share their results through presentations to some local and national conferences and through participation at the annual NSF-sponsored principal investigator meetings. In addition, ATE centers could be expected to find suitable partners for their ongoing national dissemination efforts, provide a vehicle for training and technical assistance to support their dissemination, and help disseminate the results of other NSF-funded projects in areas that relate to their designated expertise.

Improving Dissemination Practices at ATE Projects

Dissemination practices at ATE projects would improve if dissemination was (1) more targeted to specific audiences such as particular students, educators, business, and government; (2) focused on solving the interrelated problems of targeted users; (3) better integrated with other aspects of project activities such as faculty development and curriculum adaptations; (4) perceived as an ongoing strategy directly involving targeted audiences at every step and not just something that occurs after materials have been prepared; (5) a key focus of center activities with state, regional, and national partners; and (6) viewed as a joint activity with NSF and the other ATE centers.

ATE centers and major projects should examine the paradigm of their organization as an "integrated solution provider" as described in this paper. The concept of "customer relationship management" also described in this paper presents some challenging ideas for collecting and using a client database to better serve new and future clients. This would help centers/projects go beyond information sharing and lead to more concrete services and impacts on their targeted clients.

Dissemination practices are likely to improve if more attention is given to impact evaluations of what is occurring. In addition to providing simple statistics such as the number of faculty and students served, centers and major projects should report on impact or effectiveness. The dissemination perspective influences evaluation by asking for evidence of the quality of products or services to be disseminated as well as by asking about the impact on users of the products and services that are disseminated. Clearly, dissemination and evaluation cannot be treated independently. Feedback from the evaluation of dissemination efforts should be reviewed each year by the centers and NSF to allow a refocusing of efforts to better inform and improve practice in advanced technology education.

Ways for Sharing Effective Practices in Dissemination

Although some ATE centers are active in disseminating their own work, it is important for center and major project staff to share with each other and the field their best practices in this area. Panels and small group discussions at the annual PI meeting in Washington, DC, should stress sharing of successful and unsuccessful dissemination strategies. In addition to sharing best practices at the annual NSF and PI conference, smaller projects may be able to post information on an e-portal for ATE projects.

Several ATE center directors suggested that they be cross-trained to speak effectively about the materials and strategies being developed by their sister centers so that educators and industry leaders with whom they associate can learn about successes at the other centers.

Centers should also be encouraged to serve as clearinghouses for information, new curricula, and best practices in the technology fields that are being developed by colleagues in other colleges or agencies across the country.

Strategies for Evaluating Dissemination Impact

A frequent outcome of evaluation is often the increased sensitivity of educators as to what is expected in dissemination because they know that someone else is looking. What can be done to strengthen the evaluation of dissemination without producing an excessive burden on busy ATE project and center staff? There are three directions to go. One is to ask questions to help ATE staff better understand who their primary clients are for the dissemination and what information they know or could find out about their clients. Second, it would be helpful to ask evaluative questions that help ATE staff see the links between what they are doing in dissemination and other aspects of their work such as research, product development, and training. Third, to help keep a focus on dissemination impact, it would be useful to encourage ATE staff to include information regarding the user impact from at least a few of the most important products or services they provide in their evaluation follow-up studies.

REFERENCES

Altschuld, J. (1999). Evaluation plan for the dissemination center. In: *Proposal for the National Dissemination Center for Career and Technical Education*. Columbus, OH: Center on Education and Training for Employment, The Ohio State University.

Curley, R. (1999). Profiting from the relationship (Industry trend or event). *Insurance & Technology*, *24*(3), 34.

Fund for the Improvement of Postsecondary Education (2000). Dissemination of successful innovations. http://www.ed.gov/offices/OPR/FIPSE/Comp/innovations.html.

Gullickson, A., Lawrenz, F., & Keiser, N. (2000). *Status report 2: Findings from a survey of ATE projects and centers*. Kalamazoo, MI: Western Michigan University – The Evaluation Center.

Gullickson, A., Lawrenz, F., & Keiser, N. (2001). *Survey 2001: The status of ATE projects and centers*. Kalamazoo, MI: Western Michigan University – The Evaluation Center.

Hutchinson, J., & Huberman, M. (1993). Knowledge, dissemination and use in science and mathematics education: A literature review. Prepared for the Directorate of Education and Human Resources, Division of Research, Evaluation and Dissemination, National Science Foundation. http://www.nsf.ov/ubs/tis1993/sf9375.txt.

Klein, S. (1992, March). A framework for redesigning an R&D-based national education dissemination system in the United States. *Knowledge: Creation, Diffusion, Utilization*, *13*(3), 255–285.

Leung, P. (1992). Translation of knowledge into practice. In: Walcott & Associates (Eds), *NIDRR National CRP Panel Final Report*. Washington, DC: Walcott & Associates.

Louis, K. S., & Jones, L. M. (2001). *Dissemination with impact: What research suggests for practice in career and technical education*. Minneapolis: National Dissemination Center for Career and Technical Education, University of Minnesota.

Louis, K. S., & van Velzen, B. (1988). Reconsidering the theory and practice of dissemination. In: R. van den Berg & U. Hameyer (Eds), *Dissemination Reconsidered: The Demands of Implementation* (pp. 261–281). Leuven, Belgium: Acco.

NCDDR (1997). *Dissemination, utilization, and the NCDDR*. http:www.ncddr.org/du/products/dsi/index.html.

NorthWest Center for Emerging Technologies (2001). *NWCET: New designs for advanced information technology education*. http://www.her.nsf.gov/PIRS_Prs_Web/Search/RetrieveReconrd.asp?Awd_Id=9813446.

Smink, J. (1985). Vocational education research on the decline – Rx: Dissemination. *Journal of Vocational Education Research*, *10*, 1–14.

U.S. Department of Education, Office of Postsecondary Education (2000). *Application for grants under the TRIO Dissemination Partnership Program*. CFDA Number: 84.344. Washington, DC: Author.

Westbrook, J., & Boethel, M. (1997). *General characteristics of effective dissemination and utilization*. http://www.ncddr.org/du/products/characteristics.html.

MATERIALS DEVELOPMENT

Gloria Rogers

INTRODUCTION TO MATERIALS DEVELOPMENT

This issue-oriented paper was written to provide a framework for the development of high-quality curriculum materials that can serve as a guide for those who are considering developing a proposal to the ATE program and to guide NSF in the review of proposals and evaluation of funded *projects* (i.e. projects and centers). To accomplish this, a review of relevant literature and an analysis of the current ATE *projects* survey and site visit reports was made. General observations and recommendations for improvement are included in the paper.

REVIEW OF THE LITERATURE

Current ATE Program Focus on Materials

The ATE program makes two types of grants: projects and centers. Projects are generally smaller grants, shorter in duration, and more focused on one or more ATE program tracks (curriculum and educational materials development, professional development for educators, technical experiences, laboratory development, and dissemination). Centers are more comprehensive than projects and are funded for a longer duration of time. There are two types of centers: National Centers of Excellence and Regional Centers for Manufacturing or Information Technology Education. The ATE proposal solicitation states that National Centers for Excellence "must have a national impact" and typically engage in the full

Evaluating the Upgrading of Technical Courses at Two-Year Colleges:
NSF'S Advanced Technological Education Program
Advances in Program Evaluation, Volume 9, 57–78
Published by Elsevier Ltd.
ISSN: 1474-7863/doi:10.1016/S1474-7863(04)09004-0

range of activities associated with the projects. It explicitly states that Centers of Excellence "are expected to develop high-quality educational materials, courses, and curricula; to provide professional development for educators to support the utilization of these resources; and to disseminate their products through commercial publishers, journals, conferences, workshops, electronic networks, and other means."[1] In relation to educational materials, ATE guidelines indicate that Regional Centers should undertake activities that address, "academic program reform, such as using industry and skill standards and other input from industry in program development, adapting and implementing exemplary educational materials and practices developed elsewhere . . .."[2]

The convention of using the italicized *projects* will be used in this paper to refer to all funded activities (both projects and centers). The term projects (unitalicized) will be used to refer to the smaller grants, and centers will refer to the larger, comprehensive grants.

The ATE program's central goal is "producing more science and engineering technicians to meet workforce demands, and improving the technical skills and the general STEM preparation of these technicians and the educators who prepare them."[3] It is designed to promote improvement in the education of science and engineering technicians at the undergraduate and secondary school levels. Activities that are eligible for funding include, "the design and implementation of new educational materials, courses, laboratories, and curricula; the adaptation of existing exemplary educational materials, courses and curricula in new educational settings; . . . and the broad dissemination of exemplary educational materials and pedagogical strategies that have been developed and previously funded ATE awards."[4] The NSF-ATE guidelines encourage proposals that will produce educational materials that can be used beyond the grantee institution. Proposals are also encouraged that adapt and implement "high-quality" educational materials that have been developed by previously funded *projects*.

In the program description for those *projects* that have indicated that materials development was a focus, the following guidelines are given:

> Proposed activities should affect the learning environment, course content, and experience of instruction for students preparing to be science and engineering technicians. Projects often result in textbooks, laboratory experiments and manuals, software, CD-ROMs, videos, and other courseware. Such products are expected to be field-tested in diverse locations and widely disseminated through commercial publishers, conferences, workshops, electronic networks, journal articles, and other means. Educational materials and curricula that offer student innovative, high-quality learning experiences through distance education are encouraged. A project's focus may range from the adaptation of existing educational materials to the creation of entirely new ones; from a few modules at a single educational level to a comprehensive curriculum for multiple years; and from a single subject to the integration of several disciplines.[5]

The guidelines allow for a range of materials development efforts – from materials that are developed for local use only for program improvement with limited dissemination to materials that are developed for commercialization with broad dissemination. The review criteria developed by NSF for the evaluation of proposals includes criteria related to materials development. In particular, reviewers are asked to consider, "What is the potential for the project to produce widely used products that can be disseminated through commercial or other channels? Are plans for producing, marketing, and distributing these products appropriate and adequate?"[6]

Evolution of NSF-ATE Proposal Guidelines

An analysis of the NSF-ATE 1994–2001 proposal guidelines was done to determine whether or not the guidelines had changed over time with regard to the emphasis on materials development and the review criteria for proposals. It was found that the guidelines related to materials development were unchanged from 1994–1996. In 1997, the general review criteria for proposals were changed to add, "What is the potential for the project to produce widely used products through commercial or other channels?" This was the only criterion that had direct reference to the production of "products" that would include materials. In 1998, there was a change in the wording of the guidelines for projects. From 1994–1997 the guidelines stated,

> Of particular interest are projects that are designed to produce major changes and significant improvement beyond the recipient institution and which will produce materials used regionally or nationally.

In 1998 and 1999, this wording was changed to read,

> Projects must produce major changes and significant improvement beyond the recipient institution and produce materials used nationally.

This change reflects a shift from encouraging projects to requiring projects to produce materials for use beyond the local setting.

The 2000 guidelines omitted the above wording and were changed to reflect the longevity of the ATE program and the need to encourage others to adapt and implement "high-quality materials" that had already been developed under the ATE and other programs. The "adaptation and implementation" section of the guidelines encouraged projects that involved an innovative use or extension of these materials in a setting different from the one in which they were created. This included the adaptation and field-testing of existing materials. The guidelines also encouraged proposals that would take materials that were developed for use in one technical field and adapt and implement them in a different technical field.

A new category of awards was also added in 2000 that was focused on "articulation partnership." Among other things, projects funded under this category were encouraged to develop or adapt "high-quality STEM educational materials, courses, and methods for use in two-year college courses that serve prospective teachers."

The review of guidelines reflected a progression of expectations and goals of the ATE program. In the beginning, projects were encouraged to develop "high-quality" materials for dissemination. In 1998–1999, the guidelines for projects were changed to read, "must . . . produce materials used nationally." In 2000, after six years of funded *projects*, the guidelines were again changed to encourage proposals that took advantage of materials already funded under the NSF-ATE and other programs through adaptation and implementation in new settings.

Materials Development Defined

The ATE driver "materials development" is defined as the preparation, adaptation for implementation and/or testing of one or more courses, modules, process models, and/or other instructional or assessment units.[7] Materials are curriculum products that result from curriculum development processes and provide the bases for instructional decisions.[8] From this perspective, the products or materials are the artifacts of the curriculum development process and reflect the decisions made throughout the process.

Curriculum Materials, Curriculum Processes, and Educational Change

Educational materials are the primary media by which curriculum is documented and disseminated. A discussion of curriculum is essential to understand the materials development process. What defines curriculum is not easily answered. Henson cites 13 different definitions of what comprises the concept of "curriculum." All of the definitions reviewed have in common the fact that each reflects values and beliefs about the focus of the educational process and the needs being served.[9] Taba offers two definitions of curriculum:[10]

> Curriculum is a plan for learning.
>
> A curriculum usually contains a statement of aims and of specific objectives; it indicates some selection and organization of content; it either implies or manifests certain patterns of learning and teaching, whether because the objectives demand them or because the content organization requires them. Finally, it includes a program of evaluation of the outcomes.

English defines the curriculum as, "the work plan or plans developed by or for teachers to use in classrooms by which the content, scope, and sequence of that content, and to some extent the methodology of their teaching, is defined and configured."[11] His definition is derived from a historical context where the word "curriculum" was synonymous with texts. He views curriculum as a document of some sort that focuses and connects the work of the classroom teacher.

In addition to the components of curriculum mentioned in the definitions above, curriculum must also meet the needs of those it serves. One variable in curriculum development that complicates the process of development is the complex and forever changing environment within which it occurs. To be effective, the curriculum needs to be flexible to adapt to the changes in society at large, the local community, and the students themselves.[12] The inability to adjust to these changes renders the curriculum ineffective to meet the needs of those it serves.

The development of curriculum materials is an integral part of the curriculum development and educational change processes. Curriculum processes are those procedures involved in creating, using, and evaluating the curriculum that is represented in various products or materials. These materials may include syllabi, curriculum guides, courses of study, resource units, lists of goals and objectives, texts and other documents that deal with the content of the education.[13] To evaluate the products or materials generated by the curriculum process, it is important to understand the steps taken in the systematic development of curricula. If materials generated are the products of a systematic curriculum process that is validated by a model for curriculum design, the likelihood that the materials will be of high quality is greatly enhanced.

Curriculum development is also a vehicle for educational change.[14] The ATE program promotes both innovation and adaptation for improvement and is designed to effect change in the scope and quality of technician education. Developing and disseminating high-quality educational curriculum materials is one of the mechanisms funded to promote this change.[15]

Educational change as used here is consistent with the definition employed by Bridges.[16] He makes an important distinction between "change" which is external to the individual and "transition" which is an internal or a psychological process. Curriculum materials are the vehicle for transporting the curriculum to the classroom teacher and student and should be designed in such a way as to effect the desired change in instruction and learning by enhancing the likelihood that the users of the materials can transition smoothly from the old curriculum to the new.

Materials also inform the multiple stakeholders in the educational process of the focus and scope of intended educational outcomes. Stakeholders in the ATE program should be able to examine the materials produced and identify instructional processes and the knowledge and skills that the intended target group is expected to achieve through the use and/or application of the materials.

The development of a framework to evaluate materials produced under the ATE program needs to take into consideration that not all materials that are developed under this program are designed to be commercialized and broadly disseminated. However, if the funded *project* has identified "materials development" as an area of focus, there should be an expectation that they will meet the criteria stated in the ATE program guidelines referred to earlier.

To establish a framework that can be used to guide and evaluate the educational materials development process, several curriculum models were reviewed. The purpose of the curriculum models is to describe a set of logical relationships that link together relevant features of the curriculum development process. These models are useful in the planning and evaluation of the effectiveness of curriculum materials. One model is a curriculum model that was chosen from several general models that ranged from simple to complex in their organization. Oliva's curriculum model was chosen because it is comprehensive and often cited by experts.[17,18] This model is compared to the "Systematic Curriculum and Instructional Development (SCID)" model.[19] The SCID model was developed to incorporate the critical tasks needed to develop competency-based education (CBE) curriculum and instructional materials for workforce training. The SCID model is the basis for the DACUM (Developing A Curriculum) process that has been utilized in several of the ATE *projects* to analyze job or occupational skills needed for expert workers.[20] The general model and the SCID model are compared in Table 1. This table identifies the common elements essential to the curriculum development process that can inform the development of a framework for the analysis of the materials development objective of the ATE program.

To facilitate the comparison, the Olivia model was placed in the left-hand column and the SCID model was mapped across to the right column. The purpose of the SCID model is more focused on workforce training and includes more detailed steps, whereas the Olivia model has fewer steps that encompass more than one step in the SCID model. The only significant element that was not comparable in the two models was that the Olivia model contains the need for an explicit statement concerning the aims and philosophy of education (including beliefs about learning).

Each of the curriculum development models emphasizes the importance of needs analysis, specification of goals and learning objectives, curriculum design based on needs analysis and goals, development of appropriate instructional strategies, formative and summative evaluation, and improvement of the curriculum based on evaluation evidence. Each of the components of the curriculum development process should be reflected in the curriculum materials. Based on the models cited and the ATE focus on technician training, the checklist below illustrates the necessary components for the development, implementation, and improvement of effective, high-quality educational (curriculum) materials. For the development of

Table 1. Comparison of Curriculum Development Models.

Olivia Model	SCID Model
1. Statement of aims and philosophy of education, including beliefs about learning	1. Curriculum analysis
2. Specification of needs	a. Needs analysis
3. Specification of curriculum goals (long term, attainable)	b. Job analysis
4. Specification of curriculum objectives (ability of student to perform selected tasks, measurable)	c. Task verification
4. Specification of curriculum objectives (ability of student to perform selected tasks, measurable) continued	d. Selection of tasks for training
5. Organize and implement the curriculum; formulate and establish the curriculum structure	e. Standard task analysis (identification of performance steps and decisions, essential knowledge, industry standards, etc., needed to develop accurate and relevant teaching and learning materials.)
8. Choice of educational strategies or delivery systems	f. Literacy task analysis–optional (knowledge category broken down appropriate skill set – communication, mathematics, science, computer, and decision-making.)
6. Specification of instructional goals for each level and/or subject	2. Curriculum design (based on information collected in phase 1)
7. Specification of instructional objectives for each level and/or subjects	a. Determine training approach
9a. Preliminary selection of evaluation techniques	b. Development of learning objectives
10. Implementation of instructional strategies	c. Development of performance measures
9b. Final selection of evaluation techniques	d. Development of training plan
11. Evaluation of instruction	3. Instructional development
12. Evaluation of curriculum	a. Development of a competency profile (competency-based programs) or develop a curriculum guide (for traditional programs)
	b. Development of learning guides/modules (competency-based programs) or lesson plans (for traditional programs)
	c. Development of supporting media
	d. Pilot test and revise the materials
	4. Training implementation
	a. Implement the training plan (bring together resources)
	b. Conduct the training

Table 1. (*Continued*)

Olivia Model	SCID Model
	c. Conduct formative (in-course) evaluation of students and instructor performance
	d. Document training (student achievement and instructor performance)
Feedback results to improve curriculum and instruction[a]	1. Program evaluation
	a. Conduct summative evaluation
	b. Analyze and interpret information

[a] Although this step is not included in the text description of the model, the feedback loop is included in the graphical model.

high-quality materials to support the curriculum development process, materials should (Table 2):

The elements of the checklist are described below:

(1) *Reflect the results of a formal needs analysis.* Educational materials should be developed based on the needs identified by the appropriate stakeholders. In most cases, the needs are identified by the industry partners for whom the technician training is targeted. In some incidences, needs analysis might also refer to analyzing the strengths and weaknesses of the current curriculum or anticipation of future needs for technical preparation.
(2) *Be industry verified.* The context of the materials should reflect industry practice, and the competencies contained in the materials should be consistent with the needs for technicians in high technology fields and be reviewed and validated by the industry partner. The materials should anticipate the needs of industry and not only reflect industry practice. Industry shifts more quickly than materials can.
(3) *Reflect learning goals and objectives.* Learning goals and objectives should be clearly articulated and measurable.

Table 2.

1. Reflect the results of a formal needs analysis	√
2. Be industry verified	√
3. Reflect learning goals and objectives	√
4. Be developed/adapted as a part of a systematic curriculum development process	√
5. Support and identify instructional strategies including pedagogy and assessment	√
6. Be pilot and field-tested	√
7. Be continuously evaluated	√
8. Be revised based on evaluation evidence	√

(4) *Be developed/adapted as a part of the curriculum development process.* There has been a formal, systematic curriculum development process followed in the preparation of the educational materials.
(5) *Support and identify instructional strategies including pedagogy and assessment.* Educational materials should include instructional and assessment strategies that can be utilized in delivering the curriculum. The relationship among materials, pedagogy and assessment is complex and the educational materials developed need to carefully consider all three and their alignment.
(6) *Be pilot and field-tested.* Materials are validated for understanding, appropriateness for multiple users, and effectiveness in promoting learning locally and at selected test sites before publication and general dissemination.
(7) *Be continuously evaluated.* Processes are in place to evaluate materials on an on-going basis during the implementation phase.
(8) *Be revised based on evaluation and research evidence.* Formative evaluation is conducted throughout the development process, and evaluation findings are regularly used to guide creation and revisions of the materials. Additionally, research should be undertaken on how the materials are working in terms of teacher understanding, student learning. Industry should play an important role in validating the outcomes of the materials.

Although the depth of these processes may vary from those *projects* developing materials that are expected to have broad impact through dissemination and/or commercialization and *projects* developing materials primarily for local use or for use within the *project* (e.g. program improvement), the breadth of application of the processes on the checklist should be addressed in some way in all materials development.

STATE OF THE ATE PROGRAM RELATED TO MATERIALS DEVELOPMENT: SURVEY RESULTS[21]

A review of the current state of the ATE program was done by examining the results of the 2001 ATE survey developed, with input from NSF, by the Western Michigan University (WMU) evaluation project at The Evaluation Center. Eighty-one *projects* participated in the survey (70 projects and 11 centers), and 75 *projects* (64 projects, 11 centers) or 93% responded to the survey. All *projects* sampled had been active for at least one year. Ninety-six percent of the 2001 sample was also sampled in 2000. Respondents were asked to complete the "Materials Development" section of the survey if the development of materials was a focus of their *project*. Sixty-two *projects* (83%) completed the materials development section of the survey. The length of time the *project* had been in existence was not taken into consideration in the analysis other than that they all had been in existence at least one year. The materials development section of the survey focused on four dimensions: (1) type of materials developed; (2) stage of development; (3) category of usage; and (4) indicators of quality. Results of each of these will be discussed separately.

Type of Materials and Stage of Development

Respondents were asked to indicate the type of materials developed. Types of material were classified as: (1) course; (2) course adaptation for implementation; (3) module development; and (4) other. Course adaptation refers to a major revision of an existing course for implementation. Module development is a component that can be used in more than one course. The "other" classification was comprised of materials developed that were not discipline specific (e.g. cooperative learning, diversity). The stages of development were: (1) draft stage; (2) being field tested; and (3) completed. A total of 3,969 materials were reported in these three categories (not mutually exclusive). Of this total, 36% of the responses were in the "draft" category. Forty percent of the responses were in the "field test" category with the remaining 24% being in the "complete" category.

Because the "being field tested" category is not mutually exclusive from the "draft stage" category, the categories "draft stage" and "completed" were combined to give a closer estimate of the total number of materials being developed. Table 3 indicates the total numbers reported to be in draft or complete stages and the percent of contribution the centers and projects made toward the total.

Table 3. Types of Materials and Stage of Development.

	Total Reported in Draft or Complete Stage	Center % of Total	Projects % of Total
Courses	475	39	61
Course adaptation	177	36	64
Modules	1,617[a]	75	25
Other	106	4	96

[a] One center reported 720 modules in the draft stage of development. The average number of modules reported by the 9 other centers was 23. If this number were substituted for the 720 to try to correct for the outlying number, the percentage of modules in draft or complete stage would be Centers: 56% and Projects: 44%.

Category of Usage

The survey asked respondents to indicate how the materials developed were being used in terms of their target audiences. The use of the materials developed was classified in three ways: (1) local use – materials that were developed to support program improvement and were used within the *project*; (2) elsewhere – materials used at sites that were not a part of the *project*; and (3) commercially published. These classifications were not mutually exclusive. The survey results indicate that the majority of materials developed are used locally, and few have been commercially published. More than 1,700 of the materials were reported in use at least locally. If one presumes all materials developed will be used at least on a local basis, then 35% of this total was used at sites other than the *projects*, and 14% were commercially published.

In order to better understand the target audiences for the materials developed, respondents were asked to provide information for up to 5 of the most important materials they developed or were developing. Sixty-two *projects* provided one or more examples of materials development. Respondents provided information on 190 incidences of material development: 76-course development, 10-course adaptation, 72-course modules, and 32-other. Approximately 80% of the developed materials were targeted at the associate degree level, and the materials represented 17 discipline areas. Of the 190 products reported, 18% were prepared for K-12, 44% for first-year college, 36% for second-year college, and 2% for upper-level college.

Quality of Materials Development Work

The status report to NSF (2001 survey results) points out that the measures of quality chosen can serve only as proxies for evidence of quality. The survey focused on validation practices on the assumption that good practices are likely to lead to good products. The quality measures identified are consistent with elements of the materials development framework described above. The survey focused on three measures of material quality:

(1) Use of industry or other relevant standards as a guide to the development of materials.
(2) Measures of student success.
(3) Extent to which the materials were tested both for development and validation purposes.

Two items were used as indicators of use of industry or other standards as a guide to materials development:

Table 4. Frequency of Use of Industry Standards or Other Relevant Guidelines for Developing Materials. $n = 62$, 52 Projects (P), 10 Centers (C).

Practice	Used Each Time or Most Times %	Used Less Than Half the Time, Almost Never or Never %	NA %
1. Obtain verification by industry regarding alignment of materials with workforce and skill needs	79 (P) 100 (C)	10 (P) 0 (C)	11 (P) 0 (C)
2. Use applicable student-and industry-based standards or guidelines to guide materials development	88 (P) 100 (C)	6 (P) 0 (C)	6 (P) 0 (C)

(1) Verification by industry regarding alignment of materials with workforce and skill needs.
(2) Use of applicable student and industry-based standards or guidelines to guide materials development.

Respondents were asked to indicate the frequency with which they used each measure. The response categories were each time, most times, less than half the time, almost never or never, and not applicable. The results are summarized in Table 4. Based on additional data analysis, 73% of the projects and 80% of the centers reported that they used one of the two practices all the time. Only 4% of the projects reported that they never or nearly never apply such developmental practices.

Most materials are developed to enhance student learning in targeted technical areas. As indicated in the 2001 survey report, the variety of materials being developed warrant different approaches to the assessment of student learning. The assessment of student learning is important to the materials development process. Five items were developed as indicators of good assessment practice that can both document student achievement and serve as a guide to instructional processes and accountability. These items were used:

(1) Assess student success (knowledge and skills) in comparison with industry/business standards (American Electronics Association Standards, American Chemical Society Standards, etc.).
(2) Assess student success (knowledge and skills) in comparison to educational standards (STEM foundation standards, AMATYC, National Council of Teachers of Mathematics Standards (NCTM), National Research Council Science Education Standards, etc.).

Table 5. Frequency of Use of Measures of Student Success. $N = 62$, 52 Projects (P), 10 Centers (C).

Practice	Used Each Time or Most Times %	Used Less Than Half the Time, Almost Never or Never Used %	NA %
1. Assess student success (knowledge and skills) in comparison with industry/business standards	54 (P) 70 (C)	23 (P) 20 (C)	23 (P) 10 (C)
2. Assess student's success (knowledge and skills) in comparison with educational standards	50 (P) 70 (C)	32 (P) 30 (C)	18 (P) 0 (C)
3. Assess student success (knowledge and skills) in comparison with nontechnical skill standards	44 (P) 90 (C)	36 (P) 10 (C)	20 (P) 0 (C)
4. Assess student success (knowledge and skills) in comparison with other nonproject or nonparticipating students	42 (P) 40 (C)	33 (P) 60 (C)	25 (P) 0 (C)
5. Assess improvement of student performance in the workforce	34 (P) 60 (C)	34 (P) 40 (C)	32 (P) 0 (C)

(3) Assess student success (knowledge and skills) in comparison with nontechnical skill standards (e.g. SCANS).
(4) Assess student success (knowledge and skills) in comparison with other nonproject or nonparticipating students.
(5) Assess improvement of student performance in the workforce.

The survey results are summarized in Table 5. Upon further data analysis, it was found that 50% of the projects and 60% of the centers applied one or more of the identified student measures each time. At the other end of the spectrum, 40% of projects and 40% of centers made little or no use of these student assessment techniques, though they deem them applicable.

The third measure of quality is the extent to which the materials were tested for development and validation purposes. On the survey, a distinction was made between pilot testing and field-testing. Pilot-testing is defined as those methods used by developers to try out the materials to ensure that they are understood, properly employed, and learned. Field-testing is routinely done when the materials are believed to be ready for dissemination. The purpose of field-testing is to take the completed product and have others try it outside the development site. The field-testing process is designed to determine if materials are clearly understood by those not involved in the development process and whether or not the anticipated learning is taking place. Respondents were not asked to comment on the results of the testing, only whether or not they had conducted testing.

The data indicate that 80 or more of the *projects* pilot-tested or locally field-tested their materials all or most of the time. However, only 70% of the centers and 48% of the projects field-tested their materials externally each or most of the time. This may reflect the fact that many *projects* are focused locally at the site of development or within the *project* schools. It may also be related to where the *projects* are in the development process.

Respondents were asked to select one item that they had developed and indicate what they believed was the most compelling evidence for its quality. Although four of the six measures of student success referred to student knowledge and skills, an analysis of 50 written responses indicated that there was almost a total reliance on statements of satisfaction by users rather than on concrete, direct evidence of the quality of the outcomes. Seventeen respondents indicated the enthusiasm and/or interest of students, teachers, employers, and/or publishers was the most compelling evidence of quality. Eight mentioned enrollment, graduation, and/or placement rates as measures of quality. Only four respondents mentioned evidence of outcomes from field or pilot testing (2 of 10 centers and 2 of 40 projects). The sampling of comments below is indicative of the range of responses concerning evidence of quality:

> Both students and teachers have expressed an interest in the module and have shown interest in using the module when teaching this topic.
>
> Our students enter the workforce with knowledge and skills necessary to be productive workers quickly.
>
> It has received good reviews from presentations at peer review conferences.
>
> This program, which originated two years ago continues to be funded at an increased level of funding for each of the three years.
>
> Our retention rate increased from an average of about 50% from first to second semester, to over 90% with the new curriculum.
>
> The most compelling evidence for its quality is the number of publishers who would like to have this in their product line.
>
> Growing number of hits on the website.
>
> Clarity of the module and the supporting documentation attests to the quality of this module.

STATE OF THE ATE PROGRAM RELATED TO MATERIALS DEVELOPMENT: SITE VISIT REPORTS

Teams of evaluators and content/process experts visited 3 centers and 10 projects to collect in-depth information concerning the status of the ATE program. The site visit teams used common procedures and format, and each submitted a report of its findings that included the status of each of the ATE drivers (program improvement, profession development, collaboration, and materials development). Each site visit report was analyzed to identify the materials development process used. It was not always clear what part of the curriculum and materials development process was begun/completed prior to the funding of the center/project, so there was no attempt to limit the analysis to only those products produced as a result of the grant. The analysis focused on the materials development process and the content of the materials developed. In particular, the following evidence was sought:

What processes were used in the curriculum development?

- Needs analysis conducted.
- Curriculum model or architecture used.
- A review process for materials developed.
- Content experts or educational consultants involved.

Did the content of the material include:

- Industry or relevant standards?
- Well-defined, clearly understood learning objectives?
- Identification of student activities?
- Pedagogy?
- Assessment methods

A review of the site visit reports found that in some cases there was not enough information on materials development in the report to adequately analyze the process. However, the analysis did reveal several generalizations about materials development at the sites visited. It is cautioned that the absence of evidence does not necessarily indicate that the particular element was not included in the materials prepared by the site, only that there was no evidence in the report that it was included.

Needs Analysis

Evidence of a formal process used by the site to determine the needs of industry was absent in most cases. The level of industry involvement varied from making suggestions as members of advisory boards to verifying the competencies during the materials development process. For the most part, evidence would suggest that input was given AFTER the proposal was funded and not the impetus for the proposal.

Curriculum Model or Architecture Used

There was evidence of the use of a curriculum model or architecture in only one center and three of the projects. In two of the cases, these curriculum models were provided as a part of the institutional or state requirements for the approval of new courses or curricula. The local curriculum development guidelines were general in nature but systematic in their structure.

A Review Process for Materials Developed

Specific evidence was found that eight of the thirteen sites visited had some type of process of review for their materials. There was no evidence that these reviews were systematic based on standards of good practice for curriculum materials development. The nature of the reviews varied from advisory board review (some members of whom were adjunct faculty at the institution) to review by industry partners who were associated with the site. In some instances, the curriculum materials were reviewed by faculty at other sites and comments provided to the developer even though the reviewer did not implement the material in his/her own classes.

Content Experts or Educational Consultants Involved in Materials Development

Evidence was found that all the centers and six of the projects visited used content experts in the preparation of their materials. Expert participation ranged from involvement of those practicing in the field to utilizing industry-based curriculum materials as the bases for the content of course and module materials development. Only one center and one project were cited as utilizing curriculum or educational experts in the development of their materials.

In a review of the site visit reports, it was generally found that there was not enough information to evaluate the existence of content elements of the curriculum. The exception to this was in the area of the use of standards that is reported below.

Standards Included

Standards refer to any formal analysis of the competencies that are needed in the workforce. Standards can be national standards (e.g. SCANS), industry-based standards, competencies developed locally through a formal process (e.g. DACUM), or state standards for education. All of the centers and nine of the projects visited used industry or related standards in the development of their materials. One project reported that they had consulted their advisory board on establishing technician and workplace standards and that skill standards were discussed with an external evaluator. However, there wasn't any evidence that the site visitors could report that any formal standards had been incorporated into the curriculum. It should be noted that in several cases an ATE project would use standards that had been developed by another ATE project or center as the basis for their educational development efforts. Several of the sites visited adopted curriculum and materials that were developed by industry to meet specific educational needs.

GENERAL OBSERVATIONS

The following general observations about materials development are made based on the review of materials related to the program and the review of literature.

Time to Production

The development of quality materials for broad dissemination and/or publication takes time, and time is of the essence in a rapidly changing technological environment. One site reported that after several years of producing high-quality materials, they had reduced their cycle for production from one and a half years to one year per module.[22] In most cases, it is over a year before grantees are even ready to begin to produce materials. Expectations for the development of quality materials to support the training of technician must be tempered with the complex and time-consuming process of quality materials development and the expectations that the materials will be kept current with the rapidly changing industries that they support. The fact that many *projects* fell short of their production goals could be contributed to the lack of understanding of the complexity of the materials development process.

Structure of the Materials Design Team

At this point, there has been no discussion of who should comprise the *project* group charged with developing instructional materials. The "Instructional Design Symposium" was held at the ATE center, Maricopa Advanced Technology Education Center (MATEC), in February 2001. The symposium brought together personnel from the ATE centers that were involved in instructional design to discuss common challenges. One of the teams of curriculum developers formed discussed the challenges of module design.[23] One issue discussed was how best to staff and organize the writing, editing, production, and testing of modules. The result was a consensus that the production of modules should have the following elements:

- Design team that consists of at least:
  - Instructional designer.
  - Faculty content expert.
  - Industry content expert.
- An established process that includes:
  - Specific subject matter and competencies.
  - Learning activities.
  - Assessments (traditional and alternative).
- A time line that includes:
  - Development.
  - Verification of module by appropriate experts.
  - Pilot testing of modules.

The Symposium team also addressed the challenges of keeping the modules current. This was identified as a critical element in the credibility of the overall process. There was little evidence that many of the ATE grantees involved curriculum designers in the development of curriculum materials. For those *projects* that are focusing on materials development, this could be a significant help in assuring the quality of the curriculum products. Those grantees that have utilized instructional design and content experts and have produced high-quality materials can be used as models for other grantees in the development of materials.

Format and Type of Materials

In some cases, the site visit reports did not indicate the nature of the materials developed. There was evidence that materials generally consisted of various forms of syllabi, course outlines, texts for students, laboratory manuals, student guides, faculty guides, and public relations/informational materials. There wasn't ample evidence to generalize about the type of material that was being produced, but several of the sites visited had only produced course outlines or expanded syllabi for dissemination to other campuses within the *project*. Some sites had produced CD-ROMs or developed interactive web sites. Others had created a combination of text and digital media to transport the curricula.

Evaluation of Quality

Although many materials have been developed in various formats, there is no evidence that the quality of the materials has been evaluated objectively. For example, for educational software, the NSF-funded NEEDS project provides guidelines for the quality of electronic educational courseware. The guidelines are outlined in criteria for a national award for excellence, the "Premier Award for Excellence in Engineering Education Courseware."[24] These award criteria are applicable to any educational courseware and provide a useful guide for evaluating electronic delivery systems. The criteria include the expanded topics of instructional design, software design, and content.

Evaluation of Effectiveness

The use of an external evaluator was often cited as the source of the evaluation of effectiveness of the materials. However, this person was usually not on-site and wasn't involved in an on-going basis, but visited the site once or twice a year, perhaps meeting with the visiting team or advisory board. When the site reports discussed the role of these evaluators, their role was generally to assess how well the overall goals were being met and not the evaluation of educational outcomes. Often the evaluators were reported to have developed surveys and other data collection instruments for the project/center. There was little evidence that there was a well-developed effort to assess and evaluate student learning or the impact of the funded work on workforce preparation.

RECOMMENDATIONS

The following recommendations are made to promote the likelihood that funded *projects* will produce high-quality materials that will promote the goals of the ATE program:

(1) Those submitting proposals to the ATE program that have materials development as a focus should be directed to (or NSF should provide) resources identifying best practices in the development of high-quality text and digital materials.

Evidence indicates that some grantees do not realize the complexity of the materials development process. The development of high-quality materials is a time consuming, systematic process that requires resources and realistic expectations of what can be accomplished in a limited period of time. For the

ATE program to achieve its goal of producing more science and engineering technicians to meet workforce demands through "the design and implementation of new educational materials, courses, laboratories, and curricula; . . . and the broad dissemination of exemplary educational materials" there needs to be a realistic understanding on the part of the proposers of what the process entails and the resources that will be needed to accomplish it. It is not clear that many grantees have experience in writing or publishing educational materials for dissemination. In order to achieve a "high-quality" product, NSF will need to take the lead in providing standards for best practice and direct proposal writers to those standards. This could be implemented by asking one of the *projects* that have developed high-quality, exemplary processes for materials development to document and publish the process used along with some practical advice and lessons learned. The guidelines provided should be geared toward the unique challenges of trying to keep current with the rapidly changing needs of technology education. Guidelines should also reflect the depth and breadth of materials that are produced for the ATE program and the various settings and resources available that are represented by the grantees.

(2) The review criteria for funding should require that materials development proposals show an understanding of the processes required for the development of high-quality materials and that the budget and time line are realistic for the materials deliverables indicated.

There needs to be a mechanism in place to ensure that the *projects* funded have a high likelihood of success, if they are proposing to develop educational materials. Evidence should be in place that those who are submitting the proposal understand the challenges of developing high-quality materials and have identified resources needed to help in that process. It is recommended that this be an explicit part of the review process. If the proposal is of general high quality and does not contain evidence of that understanding, it is further recommended that NSF work with the principal investigator to ensure there is a systematic process in place and that resources are available to produce high-quality materials.

(3) The current reporting process of funded *projects* should be revised to include reports on the materials development processes included in the framework provided for best practice. This information will encourage accountability and reinforce the need to use best practices. It will also provide NSF with the data needed to assess the overall effectiveness of ATE funding in producing high-quality materials.

Because materials produced by the ATE *projects* are the primary vehicles for change outside the funded *project*, it is important to be able to evaluate the quality of the processes used to produce the materials. Although the use of valid processes is not a guarantee of high-quality materials, there is an implicit assumption that the use of systematic, proven processes will lead to a quality product. Given this, for each *project* that has as one of its primary goals to develop materials for broad dissemination and/or publication, it is recommended that the ATE program include in its reporting expectations that the grantee report on the processes used in development.

(4) NSF should provide workshops for ATE *projects* that have materials development as one of their primary objectives.

Workshops should be sponsored by NSF and offered to all grantees who have identified materials development as one of their deliverables. This workshop could be held annually *before* the beginning of the annual funding cycle. The workshop could be organized and facilitated by the ATE centers that have been successful in the production of high-quality materials through the use of well-defined systematic processes. This would link the work of previously funded ATE *projects* to good practices and provide examples and lessons learned that would be meaningful to grantees. It is suggested that the workshop be highly interactive and involve facilitators who represent process experts (e.g. DACUM leaders, curriculum designers) as well as the successful grantees. Examples should be provided for course, modules, and curriculum as well as various media (print, digital). These workshops would reduce that likelihood that many of the grantees will experience "false starts" or consume valuable time in seeking out appropriate resources.

NOTES

1. Advanced Technological Education (ATE) Program Solicitation, NSF 00-62, 2000, p. 6.
2. *Ibid.* p. 7.
3. *Ibid.* p. 4.
4. *Ibid.* p. 5.
5. *Ibid.* p. 5.
6. *Ibid.* p. 17.
7. *Survey* (2001). *The status of ATE projects and centers.* The Evaluation Center, Western Michigan University, Kalamazoo, Michigan.
8. Sowell, E. J. (1996). *Curriculum: An integrative introduction.* NJ: Prentice-Hall Inc., 11.
9. Henson, K. T. (2001). *Curriculum planning: Integrating multiculturalism, constructivism, and education reform* (Vol. 10). New York: McGraw-Hill.
10. Taba, H. (1962). *Curriculum development: Theory and practice.* New York: Harcourt, Brace, Jovanovich.

11. English, F. W. (2000). *Deciding what to teach and test: Developing, aligning, and auditing the curriculum* (Vol. 17). New York: Corwin Press.

12. Henson, p. 14.

13. Sowell, pp. 11–12.

14. Fullen, M. (1993). *Change forces: Probing the depths of educational reform*. Bristol, PA: Falmer Press.

15. *Advanced Technological Education (ATE) Program Solicitation* (p. 4). Washington, DC: NSF (00-62, 2000).

16. Bridges, W. (1991). *Managing transitions: Making the most of change* (p. 3). Reading, MA: Addison Wesley.

17. Oliva, P. F. (1992). *Developing the curriculum*. New York: Addison Wesley.

18. Henson *op cit*.

19. Norton, R. E. (1997). *DACUM handbook*. Columbus, OH: Center on Education and Training for Employment, College of Education, The Ohio State University.

20. In the DACUM process, a trained facilitator meets with a committee of 5-12 expert workers from the area of analysis and develops a job profile. The two-day workshop culminates in a detailed and graphic portrayal of the duties and tasks performed by the workers involved. Lists of the general knowledge and skills, worker behaviors, tools/equipment/materials/supplies, and future job trends/concerns are also identified (Norton).

21. For a complete discussion of survey results with associated tables, see *Survey* (2001). *The Status of ATE Projects and Centers*. Kalamazoo, Michigan: The Evaluation Center, Western Michigan University.

22. *Instructional Design and Curriculum Development Symposium Summary Report to the National Science Foundation*, Spring 2001.

23. *Ibid*.

24. http://www.needs.org/engineering/premier/2000/criteria.html.

PROGRAM IMPROVEMENT EFFORTS: APPROACHES AND RESULTS IN IMPROVING TECHNICAL EDUCATION PROGRAMS

Lester W. Reed Jr.

ORIGINS OF ATE PROGRAM IMPROVEMENT EFFORTS

The NSF's ATE program responds to the *Scientific and Advanced – Technology Act of 1992 (PL 102–476)*. The Act's purpose is to improve the competitiveness of the U.S. in international trade by increasing productivity through increasing the pool of skilled technicians. To accomplish this, NSF was directed to establish a national program to improve the education for technicians in advanced technology fields. Although secondary schools and four-year colleges have a role in accomplishing this objective, two-year colleges are the major focus for increasing the pool of skilled technicians. NSF, to carry out the Congressional mandate, initiated the ATE program.[1] In turn, ATE established the development of model instructional programs in advanced technology fields as a major objective.[2]

For the purpose of this paper, many ATE-funded activities associated with developing model instructional programs are identified as *program improvement*. The WMU evaluation project[3] views this as a process of comprehensive curriculum development and associated improvements that results in the production of credentialed, cutting-edge, skilled technicians. Comprehensive curriculum

Evaluating the Upgrading of Technical Courses at Two-Year Colleges: NSF'S Advanced Technological Education Program
Advances in Program Evaluation, Volume 9, 79–104
Published by Elsevier Ltd.
ISSN: 1474-7863/doi:10.1016/S1474-7863(04)09005-2

development and associated improvement are efforts that address changes to the multiple aspects of a technical degree or other technical award program. These include identifying and integrating industry standards and workplace competencies; creating instructional module/course changes; adding rigorous STEM content; implementing work-based education components; facilitating equipment modernization; integrating appropriate pedagogical approaches; increasing minority participation; providing faculty development; etc.

Activities listed above that are directly associated with course/instructional development and improvement is referred to by WMU evaluation project as *Materials Development*. Materials development can also be a stand-alone activity not directly tied to a specific academic program. However, when pursued as part of an integrated effort to develop new or upgrade existing technical programs, materials development is a major component in *program improvement*. The major difference between program improvement and materials development is that ATE *projects* involved in program improvement are engaged in materials development as well as other comprehensive activities such as standards development, faculty professional development, and recruitment/retention activities; and the program results in the award of an appropriate degree or certificate of accomplishment to program completers.

PROGRAM IMPROVEMENT DEFINED

For the purpose of this paper, a hypothetical construct for *program improvement* is needed to provide a framework for discussing and evaluating ATE's efforts to create model instructional programs. The approach selected is to define the elements that are included in program improvement efforts. The components of the definition to be used in the paper are presented below.

Elements of Program Improvement

The term program is perhaps the key to developing a workable construct. In terms of the ATE effort, program has many meanings depending on the context in which it is used. For the purpose of this paper, a program will: (1) be educational in nature; (2) generally be led by a community college;[4] (3) focus on preparing students in advanced technological fields; (4) be an organized sequence of classes; (5) consist of a comprehensive curriculum spanning an extended period of time; and (6) result in increased numbers of credentialed, skilled technicians in the workforce.[5]

The construct "improvement" also is multifaceted and includes the following elements: (1) implements the national science, mathematics, technology, and industry standards in education; (2) creates effective courses in science and technologies based on strong student mathematics and science backgrounds; (3) recognizes current and projected occupational requirements and gives prospective technicians insight into real-world work environments; (4) serves first-time and returning students and workers seeking new career opportunities or skills; (5) provides students experience with appropriate equipment; (6) integrates instructional approaches that encourages student writing, oral presentations, group learning, and completing long term projects; and (7) engages students in the practice and thought processes of formulating problems and questions, designing appropriate models, troubleshooting, and using technological tools.[6]

ATE also provides some additional expectations for comprehensive curriculum improvement efforts that pertain to the concept of *program improvement* outlined above. The first of these is *articulation.*[7] The expectation is that improved curriculum will be articulated between secondary schools, two-year colleges, and four-year colleges and universities. The second is *achievement of workforce diversity* by recruiting, retaining, [educating] and placing students from groups underrepresented in STEM fields.[8] These expectations impact program improvement efforts, particularly in areas of content (ensuring acceptability at higher levels), and the use of a variety of pedagogical approaches appropriate to diverse learning styles. To meet both of these expectations, the availability of a series of developmental or bridging courses to attain the readiness needed to master the curriculum can be required.[9]

One other source that delineates program improvement is the WMU evaluation project. This project defines program improvement as "*efforts to create effective workforce technician education programs at the (a) secondary school, (b) associate degree, and (c) baccalaureate degree levels. 'Program improvement' refers to multiple, related courses, and/or field experiences for students at the designated education level that facilitates attaining or leads to a defined outcome such as a degree, certification, or occupational completion point.*"[10]

The WMU evaluation project's definition adds three parameters to the construct of *program improvement* that will be used in this paper. Based on this definition, there are three distinct academic levels for improvement – secondary schools, associate degree, and baccalaureate degree. Additionally, the definition stipulates the program's end result is a defined outcome, that is, leads to an award (such as an academic degree) or attainment of a recognized occupational level (The author assumes that the definition implies that the attainment of an occupational level is associated with a formally recognized standard). WMU's definition also includes the components of an organized sequence of classes that form a comprehensive curriculum spanning an extended period of time.

The Program Improvement Construct

Using the information provided above, the construct for program improvement used in this paper is ATE-sponsored program improvement is a curriculum[11] development and implementation process, normally led by a community college, which revises or develops an educational program that prepares diverse student populations with the knowledge and skills required for employment in a specific advanced-technological field. The program is an organized sequence of classes, laboratories, and work-based educational experiences available to students over a significant period of time and emphasizes STEM standards, communication skills, critical thinking, advanced technology courses, workplace competencies, equipment use, teamwork and perseverance. The improved program leads students to an appropriate degree, certification, or occupational competency point, and provides industry with an increased pool of competent skilled-technicians. Programs can be at secondary schools, two-year colleges, or four-year colleges or universities and should be structured to obtain maximum articulation of educational experiences.

Elements Used to Judge Program Improvement

The above construct is lengthy and, in some ways, a complex melding of components and outcomes. However, a simpler definition would not provide an adequate framework for this paper's exploration of ATE centers'/projects'[12] performance in improving their curriculum programs. When discussing the data about *project*s with a stated objective of producing skilled technicians, the following definitional elements from the program improvement definition will be judged. The *project*:

- Revises or develops a technical educational program.
- Reflects an integrated curriculum development and implementation process.
- Is an organized sequence of classes, laboratories, and work-based educational experiences.
- Emphasizes STEM standards, communication skills, critical thinking, advanced technology courses, workplace competencies, equipment use, teamwork, and perseverance.
- Is conducted and available to students over a significant period of time.
- Prepares diverse student populations with the knowledge and skills required for employment in a specific advanced technological field.

- Leads students to an appropriate degree, certification, or occupational competency point.
- Provides industry with an increased pool of competent skilled technicians.
- Is structured to obtain maximum articulation of educational experiences.

THE PROGRAM DEVELOPMENT PROCESS

Prior to reviewing these performance data, a limited overview of how technical programs are developed is of value in creating a perspective of what ATE-funded *projects* would be expected to do during their efforts to improve a program. Although it might seem that the process would be different, depending on whether the program was being "revised" or "created," in actuality the difference is more in degree than activities. Both program revision and creation start with a need. The specifics of how the need is determined and quantified is relatively unimportant as long as it is factually sound. It is probably safe to assume that a proposal for program improvement (as described above) submitted to ATE for funding consideration clearly establishes the need in a context of providing an increase of skilled technicians for the U.S.'s workforce. If not, it could reasonably be expected that the proposal would not be funded. Therefore, as a starting point, this paper will be based on the assumption that funded program improvement *projects* meet a recognizable manpower development need.

Differing Levels of Program Improvement

As defined, program improvement can occur at one or more educational levels – secondary school, two-year college, or four-year college and university levels. Based on the concept that community colleges will provide leadership in ATE-funded *projects*, the most common effort would be expected at the two-year college level.

The two-year college generally has a variety of "levels of occupational attainment." These consist of programs ranging from approximately one academic year with no or limited related "general education courses"[13] to the associate degree consisting of 60 or more semester hours of study, including an array of general and technical education courses.[14] In some colleges, there are intermediate levels that include a minimum of general education and a more limited number of technical offerings than in the degree program. A variation on the theme is noncredit education programs designed for workers upgrading their skill and knowledge levels and consisting of sequential technical offerings. In terms of program offerings, the community college, to serve its multifaceted enrollments, normally provides multiple options for potentially diverse populations of technical students. Therefore, it is important to clearly identify the audience when reviewing program improvement efforts at the two-year college level. An example would be the "all technical content" sequence for "reverse transfer students"[15] and the associate degree sequence for students with no or little postsecondary education experience. Both of these would be considered technical programs in context of the ATE program and this paper.

The matrix of occupational attainment levels is less complex at the secondary school and the four-year level. Secondary schools, with the exception of vocational programs, tend to have a core of mandated curriculum offerings that are general knowledge/skill building vs. intense preparation in a specific technical field. However, the educational reform movement sweeping the U.S. along with the federal school-to-work initiative have increased the focus of K-12 education toward preparing students for future employment, including increasing technological awareness and use. These efforts, however, normally do not provide concentrated, discipline-oriented instruction in advanced technical fields. The vocational education component of secondary schools does aim to provide high school graduates with employability skills without further education. These vocational courses, considering they represent the 11th and 12th grade levels of secondary education, may not contain the intense emphasis on STEM-based technology desired by ATE programs. ATE program guidance appears to recognize this limitation and speaks to collaboration with secondary schools in the design of curricula and instructional materials that provide a foundation for technical education.[16] There are articulated vocational programs at community colleges that provide advance standing for students in matching disciplines.

At senior institutions, the programs are oriented toward the baccalaureate degree or higher. At the baccalaureate level, numerous disciplines are advanced technology oriented and contain an emphasis on STEM-driven courses. ATE program solicitation guidelines take special note of the direct relationship of these baccalaureate degrees to associate degree technical disciplines. The solicitation guidelines allow support of "partnerships in which two-year colleges work with four-year colleges and universities to develop, implement, and evaluate model programs that enable students to make successful transition from a STEM associate's degree program to a related bachelors degree program."[17]

The matrix of educational levels that meet the program improvement definition used in this paper consists of a variety of community college offerings from noncredit professional development programs to multiple-year associate degree programs. At the four-year level, the bachelor's degree is the primary award. The achievement of a defined occupational level in secondary schools is generally restricted to vocational tracks during the 11–12th grade years.

External Requirements and Program Improvement

In this section, the general parameters that are usually imposed on new curriculum/program development are discussed. Since each institution, state, and region has its own rules, there will be no attempt to provide specific requirements, but a general pattern of reviews and approvals will be presented. However, these procedures are universal for state-funded institutions and vary only in degree of complexity and control of detail.

Several levels of governance/management impact new program improvement. A high school, two-or four-year college, or a university cannot simply say "let's develop program x" and do it. The process for approval of a college's degree program, particularly at publicly supported institutions, is lengthy. The first level of approvals is the institution itself. There are administrative processes, faculty review, and usually institutional governance approval for new programs.

For two-and four-year degree programs, there is a state level process for approval of new programs (and sometimes even for new courses). The rigor of the review varies greatly. In the most centralized systems, a "state approved standardized curriculum" is required. In others, the approval process is a review of a more limited proposal. The author's experience indicates a trend toward more control by states as public funding dollars have become less available.

The processes highlighted above pertain mainly to new degree offerings (and in some cases formal certificate programs below the associate degree level). Time for approval of new programs can range from one to two years. Revising existing programs is usually easier to accomplish from a review and approval process standpoint. However, as newer technologies emerge, new programs will require development. A method of "fast tracking" approval of new technical programs needed by business/industry could be beneficial in meeting needs for newer disciplines such as those in biotechnology.

Additionally, there are usually state requirements for lay business/industry advisory committees at the associate level to continuously work with and advise technical program faculty. These committees may be active in reviewing and recommending modification to curriculum content.

Regional accreditation agencies also impose requirements on formal degree programs. These requirements are generally in the mix of general education and technical courses and credentialing of faculty teaching the program's courses.

At both two-and four-year degree levels, technical degree programs are often accredited by a professional organization. To achieve accreditation (desirable in terms of external credibility) programs have to meet the organization's standards and undergo periodic review.

At the secondary school level, there are usually state-approved curricula that set minimum course requirements for graduating. Vocational programs are also controlled at the state level, although schools have a great deal of flexibility in specific content. In nonvocational programs, instructional design, specific content, laboratory modules, and educational materials are often controlled at the school level and, in many situations, at the individual teacher level.

THE STATE OF ATE PROGRAM IMPROVEMENT EFFORTS

To what degree have ATE *projects* succeeded in program improvement efforts? This is a legitimate question, considering that program improvement embodies the *Scientific and Advanced – Technology Act of 1992s* ultimate purpose of creating a world-class technical workforce. A review of several sources of information is presented below that focuses on the state of ATE's program improvement efforts and the degree these efforts have been successful.

- Surveys of ATE *projects* conducted by the WMU evaluation project (2000 and 2001 editions).[18]
- Reports of site visits to selected *projects* conducted by the WMU evaluation project.[19]
- A review of sample program course outlines and descriptions of improved programs.[20]

These data sources are analyzed, summarized, and presented in the context of the nine key elements of ATE program improvement identified above (The surveys gathered data on three levels of program improvement – secondary, associate degree, and baccalaureate. In keeping with this approach, the information provided in this section will, to the extent feasible, report on the same three levels of effort).

Revises or Develops a Technical Educational Program

The data reported by *projects* reveals that the vast majority of the improvement efforts are focused at the associate degree institutions (a reflection of Congressional and NSF-ATE intent reported above). Of the *projects* responding to the surveys, 93% reported efforts were at the associate degree level.[21] Of the 57 *projects* reporting improvement programs, 29 were exclusively at the associate degree level, while two were at the secondary level only. None of the reported *projects* focused only at the baccalaureate level. The remaining 26 activities were a combination, with all but one involving the associate degree level.

Three hundred seventy-one programs were reported as "developed/offered" by the 57 *projects*. The majority of reported programs were at a combination award level (197), 168 were exclusively associate level, 6 were at the secondary level, and none were at the baccalaureate level. These data are fairly consistent in the 2001 report.

Seven site visit reports deal with associate-degree college *projects* with a focus on program improvement as defined above. These programs cover a variety of disciplines and some have several award levels ranging from a limited and/or advanced-certificate to an associate degree. Data from review of program documentation were consistent with the data from the surveys and site visits, since all selected programs were at the associate degree level institutions.

Based on these data, it is clear ATE program improvement *projects* are developing or revising one or more specific technical program(s). For a variety of factors (e.g. previous identified need, approval process, existing expertise, and faculty), *projects* show a distinct preference for program revision.

Reflects an Integrated Curriculum Development and Implementation Process

Although the available data on how the *projects* developed and implemented their improved programs are limited, based on site visit report comments, it appears the *projects* approach the task in a structured manner. Most *projects* were in collaboration with other two-year colleges, and in many instances, the program improvement process was decentralized. Partners in these consortia-like arrangements were actually developing and implementing products, and in most cases, were responsible for program improvement efforts at their colleges. These distributive approaches to program improvement are in keeping with ATE's objective of impacting the greatest number of programs regionally or nationally. In terms of evaluation, the consortium work could not be directly reviewed and relied on the report of the *project's* focal point personnel.

Is an Organized Sequence of Classes, Laboratories, and Work-Based Educational Experiences

Those programs for which data were available presented a sequence of learning experiences that seemed logical and used active learning strategies.[22] Based on course descriptions, the technical course content makes extensive use of laboratory, field, and work-based learning/teaching strategies. The program outlines and course descriptions reflect an integration of knowledge and skill-building activities and an emphasis on applying the knowledge and skills throughout the curriculum. Most programs examined had a "capstone course" at the end of the program that required students to demonstrate an ability to integrate their knowledge and skills. However, the issue of a sequenced curriculum of building blocks of knowledge, skills, and competencies that led to qualification as a skilled technician was not directly investigated. For optimum learning, the sequencing of learning experience that reinforces and applies previous learning should be a basic approach in improved programs.[23]

Emphasizes STEM Standards, Communication Skills, Critical Thinking, Advanced Technology Courses, Workplace Competencies, Equipment Use, Teamwork, and Perseverance

The survey requested that respondents choose one specific program in one location and report on that program in some detail to minimize burden on the respondents. The *projects* reporting program improvement activities reported on the type of course development attributable to grant funding in their programs. These data were not broken down by educational level but displayed by content and action. Developing new courses and revising current courses were essentially tied in terms of numbers in 2000. In 2001, a notable increase in new courses was reported. Since developing new courses requires more time than revising an existing course, it seems reasonable that the 2001 increase in new courses can be attributed to *project* maturity. The content areas of new/revised courses consisted primarily of STEM, field-related (workforce based in nonfield settings), and technology topics. The topic areas also seem reasonable for course improvement in STEM programs.

Site visit *projects* indicated that the major objective of course/materials development is strengthening science and mathematics as well as integrating advanced technology content. Use of industry workplace standards and competencies are universally reported as the bases for program improvement. Use of DACUMs,[24] industry standards (particularly in the information technology [IT] area), and meetings with advisory committees were the reported source of these standards. Some reports alluded to the inclusion of soft skills (e.g. teamwork, communications, critical thinking) in improved programs. Work-based learning, such as internships, is available in all programs but not required in all. One *project* stated that although preferred, mandatory internships were not possible due to limited positions in industry.

All the reviewed documentation indicated that degree programs require one or more core mathematics courses in their ATE program (One course is normally a minimum required in any associate degree program by regional accreditation associations). The content of these courses varied widely. Typical approaches to the core mathematics included those listed below:

- A transfer level statistics course for which an equivalent readiness of intermediate algebra is required.[25]
- A transfer level plane trigonometry course for which an equivalent readiness of intermediate algebra is required.
- Intermediate algebra or advanced/college algebra.
- Both a trigonometry and algebra course.
- Technical mathematics.[26]

A review of the program documentation does not show any requirements for mathematics above statistics or college algebra. Based on these data, one can assume that the mathematics required in technical courses does not exceed the required curriculum level. Also, the mathematics required by associate degree technical programs (for which course descriptions were available) is generally below that in four-year programs, but is at an acceptable level for the competencies needed in the "skilled technician" workforce.[27] Catalogs and program displays do, however, caution students that higher-level mathematics course are advisable, if the student plans to transfer to a four-year program.

Although quite variable in level and content, integrating the use of mathematics in technical offerings was apparent in many technical course descriptions. Reinforcement of mathematics skills via application throughout the curriculum seemed to be a fairly consistent.

In the area of science, the variation was considerably more than in mathematics. This is predictable because of the differing disciplines of the ATE-improved programs and the relevant science. In programs with life science disciplines, the science requirements were usually in the area of biology, and the technical components were heavily weighted toward use of scientific procedures and information. The same was true in environmental disciplines, although emerging environmental science courses were also evident. Chemistry and/or physics was also required in some programs, usually coupled with advanced/college algebra. Several program documents reflected no requirement for a general education course in science. In some of these cases, the use of science methodology was reflected in technical course descriptions. In reviewing technical course descriptions, the extensive use of "inquiry based laboratory and field based modules" was evident.

The one exception was in the Information Technology (IT) area. Here the science requirement was limited (or nonexistent) and the technical courses were primarily computer based. The nature of the IT discipline (using, networking, repairing, and programming computers) and the heavy use of industry-developed instructional materials designed for certification appeared to limit the traditional science requirements. The analytical thought process associated with science seemed to be reflected in some technical course descriptions; however, many appeared to be phrased in "learn to do" terminology.

A curriculum strong point seemed to be communications skills. General education courses were uniformly at the transfer level and included writing and oral presentation development. In all disciplines, many technical course descriptions contained a theme of students presenting analytical information in writing or orally. Based on catalog information, it was evident there was adequate emphasis on communication skills.

Several other items of interest in evaluating program improvement were distilled from the course descriptions:

(1) There is an emphasis on critical thinking skills.
(2) Working in teams is used in numerous courses in an effort to develop this skill.
(3) Many programs have a capstone course that requires students to demonstrate the integrated use of knowledge and skills developed during the program.
(4) Advanced technologies and use of modern equipment associated with a particular discipline are included in programs.
(5) Programs use actual or simulated work-based educational experiences as part of their curriculum.
(6) Students are expected to be active versus passive learners in their technical offerings.

The above discussion deals with programs leading to the associate degree. Documents and site visit reports also identify credit and noncredit certificate options for many of these programs. As a general rule, certificate programs include little or no general education.

However, the technical content is the same.[28] The intended audience for these courses includes "reverse transfer" students,[29] upgrading professionals, and students seeking employment skills in a minimum of time. Except for the students' who defer general education and take technical offerings only, these certificate offerings seem very appropriate. For those students with limited mathematics, science, and communications readiness, achieving the competencies expected from the technical offerings can present a challenge.

Of course, the unanswered question is "What is the impact of this effort on student learning?" Currently, there is no discernable data about the efficacy of the improvements being made. In general, these data will be difficult to obtain and verify independently, since collection methodologies are not in place at all projects. However, to fully address the impact of program improvement, more data are needed on the following:

- *Pilot and field-testing of instructional materials.* Except in one case where state-approved new curriculum requires pilot testing, there is no mention of the process of verification of the effectiveness of materials/programs in site visit reports, nor are plans to do so presented. In one report, the lack of such testing was cited as a potential weakness. No attempts or plans to gather longitudinal field-testing data were reported.
- *Verification of industry standards.* Based on survey data and site visits, there was evidence of the initial development and use of industry standards in program improvement *projects.* Many *projects* had not developed an industry-based verification process as these standards were integrated into courses/materials of the program. Nor was there evidence of a systematic way to update the standards except through program advisory comments. The exceptions were standards developed by the Information Technology (IT) industry and used as criteria for formal certification.
- *Documentation of program materials.* Generally, comments in site visit reports indicate limited and incomplete documentation of newly developed or revised courses/modules. Since many of the program improvement *projects* are consortia or partnership based, the lack of adequate documentation may have dramatically reduced use of materials by other collaborating programs as well as verification of their effectiveness. Lack of or poor documentations would also constrain dissemination of the ATE product.
- *Pedagogical issues.* Use of appropriate pedagogy is not generally addressed, but the site visit comments that do appear express concerns that this is not a strong aspect of program improvement at the visited *projects.* At issue is whether *projects* are infusing a variety of teaching styles required by a diverse student population.[30]
- *Extent that science and mathematics has been enhanced.* From both the survey and site visit reports, it is clear that *projects* are attempting to increase the amount and rigor of science and mathematics in improved technical programs, both in core or general education as well as technical courses.[31] However, neither source addresses the degree to which these efforts are successful.

Overall, the program improvement process was clearly focused on creating curricula that are more rigorous in STEM and communication skills. Programs were designed around workforce competencies and reflected use of appropriate equipment. Curriculum design featured application of mathematics, science, and communication skills. Programs required students to work in teams and created scenarios that required students to solve problems by using critical thinking and to provide their rationale for the problem solution proposed. These problem-solving aspects of the programs are generally based on use of research skills, including those associated with the Internet. As mentioned above, it should be noted that verification of the program's content is constrained by lack of course documentation. The main issue, however, is not the intent or design of the improved programs, but how effective these efforts are in producing skilled technicians. For example, data indicate there is a minimum use of business and industry to confirm standards or evaluate the results of a curriculum. There is no evidence of extensive field-testing of programs, and *projects* do not have plans for a longitudinal study of completers' performance.

Is Conducted and Available to Students Over a Significant Period of Time

Based on information gathered at site visits and from program documents, programs range from less than a year to approximately two years. The two-year programs are both secondary and associate degree based. Certificates at community colleges can be earned in one to three semesters. No examples of four-year programs were available, but based on an assumption such programs would include transfer from an associate degree program, it is anticipated that they would consist of two or more years of study.

Prepares Diverse Student Populations With The Knowledge And Skills Required For Employment In A Specific Advanced-Technological Field

As defined, one objective of ATE program improvement is increased diversity. Recruitment of diverse students is, therefore, an expected activity if funded projects are to make a significant impact on increasing the number of underrepresented populations[32] in the technical workforce. Based on survey and site visit data, the dedication by ATE programs to recruiting is variable, ranging from none to formalized plans and targeted activities. At secondary schools, recruitment appears to be focused on middle school students with the objective of channeling them into technical offerings. The recruitment effort focus of associate level programs is both outside and inside the institution. Recruitment outside the institution appears to be more of an institution-wide effort and involves ATE projects in collaboration

Table 1.

Student Descriptor	Secondary Level (%)	Associate Level (%)	Baccalaureate Level (%)
Female	20	29	26
Minority	52	40	52
White	43	42	27
Disabilities	7	2	8

with special student services/success units. This associate degree institution-wide approach seems appropriate, since it focuses greater resources on the issue of recruitment. ATE funds can be used more effectively to enhance the institutional effort than to create a separate effort. Baccalaureate recruiting is generally focused outside the institution. Overall, the data did not provide a uniformly clear focus by projects on increasing underrepresented population participation in ATE-funded programs. Nor did these data reflect the degree of success, except for anecdotal comments, such as "increased female sign-up," "Hard to assess," "Excellent," and "No response."

However, diversity data were reported in the 2000 and 2001 surveys for both projects and centers. The data for 2001 are presented in[33] Table 1.

Although these data do not shed much light on the issue of ATE *projects* actually "creating a more diverse workforce," the number of minorities in programs seems to indicate that there is significant diversity based on race and ethnicity.[34] Based on the author's observations at ATE sites, the reported female participation appears low.

As discussed above, use of appropriate pedagogy is not generally addressed in detail. At issue is whether *projects* are infusing a variety of teaching styles required by a diverse student population.[35]

Leads Students to an Appropriate Degree, Certification, or Occupational Competency Point

The concentration of program improvement program award levels is, as expected, at the associate degree institution. The program award/educational level matrix (programs of the n programs per degree level that offer degrees or certificates [e.g. 34/36 offer a degree at the associate level) for 2000 and 2001 is reflected in Table 2.

The above distribution raises a critical question in the mind of the author. How can secondary schools grant a degree? The assumption offered is that the secondary school *projects* are in collaboration with a degree-granting institution (most likely a community college), and the award is made at the college level,

Table 2.

Award Level	Secondary ($n = 18$)		Associate ($n = 36$)		Baccalaureate ($n = 5$)	
Degree	***5***	6	***34***	34	***5***	3
Certificate	***7***	12	***21***	21	***2***	0

Note: Nonitalicized numbers are 2000 data, ***bolded italicized numbers*** 2001 data.

not the secondary level. The reported data also reflect the award of certificates at the baccalaureate level in 2000 data, which again seems interesting, since it is not normal in these institutions to do so, except in non-credit continuing education offerings. Perhaps future survey instructions might require more specific guidance on how these data are to be reported.

Based on the survey data, the site visit reports, and reviewed documentation, all the reported programs appeared to provide appropriate credentialing. The reported articulation agreements and the numbers of students "continuing their education" clearly indicate there is a demand for higher degrees in the advanced technology field (Data presented below indicate approximately one-third of program completers continue their education).

Provides Industry with an Increased Pool of Competent Skilled Technicians

Based on survey results, the WMU evaluation project's Survey Status Report concludes, "Large numbers of students are being impacted by the reported improved programs." This conclusion is based, in part, on data reported on new/improved course enrollments that show average enrollments of 76 at the secondary level and 1,075 at the associate level in 2000. In the 2001 survey, these numbers ballooned to 700 secondary and 2,300 associate degree enrollees. Based on how the item in the survey was phrased, the reported data would not necessarily reflect technical program majors. *Projects* report unduplicated head counts of students taking one or more improved courses regardless of major. Additionally, these 2001 reported average enrollments were significantly skewed to the high side by reported data at the secondary level of 2,000 and 5,000 enrollees in 2 secondary programs and 1 center reporting an enrollment of 70,000.[36] Different statistical techniques could have been employed to reduce the bias created by the several exceptionally high reported enrollments.[37]

The survey did ask for data on program enrollments and completers for the special reported programs.[38] These data are directly tied to the number of students potentially moving into the skilled technician workforce. Again, it should be noted these data are based on a single program per *project* selected by the *project* leaders and not a structured or controlled sample. The numbers presented over the 12-month periods follow (none reported at the four-year level):

Secondary:

- 2000 – An average of 173 enrollees with 125 completers.
- 2001 – An average of 98 enrollees with 40 completers.

Associate:

- 2000 – An average of 94 enrollees with 43 completers.
- 2001 – An average of 160 enrollees with 58 completers.

2000 survey data concerning program completers at the associate degree level indicate approximately 75% of the completers entered the workforce and 33% were pursuing higher education.[39] The 2001 data reflect a decrease in completers entering technical positions (46%). This reduction of completers entering the workforce was not the result of greater numbers of "continuing education," which remained around 30% for both years. Whether the drop in the proportion of students placed in technician positions was due to the recent economic downturn in the technology sector or another factor (e.g. more students in the first year of a 2-year degree program in 2001 than in 2000) remains to be determined. Additionally, the data swings may be based on the lack of data at the *project* level. Based on the author's experience, tracking of program graduates is not a universally high priority at community colleges.

The above data indicate a growth in associate degree programs with some drop-off in secondary programs. With the increasing ATE emphasis on associate-degree-based programs, this change seems reasonable. However, based on the reported increased secondary course enrollment noted above (624 more in 2001 than 2000), the various data elements dealing with secondary enrollment/completion numbers are hard to reconcile. Refinement in how these data are gathered/displayed to distinguish secondary students enrolled in a technology course(s) vs. those in a vocational/technical program would provide a clearer picture of ATE enrollments.

Based on site visit reports, the production of skilled workers from ATE improved programs is currently limited. Classes visited were small, and larger ones often included nonmajors, which could indicate that institutionalization is occurring. For example, one site reported 250 program enrollees, but in reality the number of majors was 13. The remaining students were in various programs and taking the IT course as a general science option. Additionally, multiple majors often required common courses, creating a larger course enrollment but with small enrollments in the several program options. However, the enrollment trends in ATE programs generally were reported to be increasing, and at some locations demand was exceeding available program seats. Particularly in the IT field, growth was constrained by lack of qualified instructors. Expanding instructor resources by using industry technicians was reported as a major focus in many of these *projects*.

Increases in completers can be expected as the ATE program matures. Data from the 2001 survey indicate that the enrollments and completers in selected associate degree programs are higher than those drawn from the 2000 survey (average enrollment 160, average completers 58 per reported program). The 2001 program completion data show a significant increase and indicate that ATE is having a growing positive impact on the U.S.'s workforce. Of course, in consortium arrangements, there are multiple production points, and total production for the *project* is larger than any single program. Also, no data are available on production from credit and noncredit certificate programs. It is probably safe to assume the number of completers from these programs is similar to credit programs and add to workforce improvement, especially by retaining existing technicians.

The ATE staff, with some assistance from the WMU evaluation project, has tried to use the survey enrollment and completion statistics to extrapolate total ATE associate degree enrollment data. The approach was to multiply the "average number of students taking at least one ATE-impacted course in the past 12 months" (2,304) by the number of "reporting *projects*" (47) to yield an "impact estimate." As a result, the staff estimated 108,000 associate degree program enrollees (The WMU evaluation project also provided median numbers [e.g. median for associate degree program enrollees was 150]). Even if the extrapolated numbers were accepted, the impact on the U.S. technical workforce would be small.[40]

Expansion of ATE's impact on the technical workforce will require a significant effort to disseminate curriculum products from the "model instructional programs." ATE is addressing this with special dissemination grants for dissemination focal points and directing special efforts to develop regional centers focused on upgrading critical technical programs in fields such as manufacturing and information technology.[41] Grants in these areas should assist the spread of improved and new course materials to additional technical programs. To measure the impact on the workforce, it will be necessary for ATE to develop and implement methodology that provides empirical data on the spread of improved technical education via dissemination and regional consortiums.[42]

Is Structured to Obtain Maximum Articulation of Educational Experiences

Another aspect of program improvement reported in the WMU evaluation project's survey is credit transferability between institutions. This is a significant issue in technical education, since historically there been a resistance by institutions to accept credits earned in technical courses as part of degree programs at a higher educational level. The articulation of student "learning experiences"[43] between institutions is a major focus of the ATE program as it pertains to workforce development. The reported data reflect that transfers between like-level institutions are more common than between differing levels. However, the transfer between associate degree and four-year programs is also a strong component, with all but three projects reporting some degree of transferability. It is clear that there is a concerted effort to achieve program articulation, particularly at the associate/associate and associate/baccalaureate degree levels.[44]

Site visit reports confirm the emphasis on associate to baccalaureate program articulation. All reported associate degree *projects* involved in program improvement had achieving and maintaining articulation with four-year programs as a major objective. Comments in the reports indicate a high degree of success by *projects* in this effort. At one location, a major state university actually developed an undergraduate and a graduate program sequence to accommodate graduates from associate degree institutions using ATE-developed programs.

The amount of credit transfer varies from program to program and is influenced primarily by the courses in the associate program. The positive impact of the increased rigor of ATE programs is evident once programs are "improved" by the number of programs negotiating new or revised articulation agreements.

Most associate degree programs are linked to secondary school improvement strategies focused on preparing high school students for community college programs in specific technical disciplines. The site visit reports indicate only one case where this relationship resulted in advanced standing of secondary students in the associate degree program. This an area where greater emphasis may be needed if secondary school vocational/technical students are to be encourage to continue their education.

SUMMARY

Basically, the program improvement *projects* are clearly meeting the developmental aspects of the ATE program. Overall, the *projects* involved in program improvement are oriented to improvement of STEM both in their programs, and in the case of associate-degree programs, in feeder secondary programs. The improved programs reflect use of mathematics, science, and communications across the technical curriculum component. Use of standards to determine student competencies is universal; and when the work-based standards are not available, programs work with business and industry to identify them. The use of work-based education as part of the curriculum is also standard, but is not always a requirement. *Projects* integrate developing "soft skills" such as teamwork and critical thinking into their programs. The improved programs provide certification or award a degree where appropriate. At the associate level, there are usually multiple educational tracks to meet the needs of students with a variety of backgrounds and goals. Articulation of transfer of credits or prior learning experiences between community colleges and their four-year counterparts is the norm. However, there is very limited articulation involving granting advanced standing of students between secondary and community college programs.

The issue to be addressed by the ATE program is not the development of the programs, but the lack of data on outcomes. Although some outcomes data for programs are gathered via surveys, these data are incomplete. Data on performance by program completers (either on the job or in continuing educations) are not routinely gathered or analyzed, nor are plans to do so evident. Additionally, the absence of formalized pilot and field-testing involving business and industry constrains verification of program effectiveness and revision to meet the original or new program/course objectives. Incomplete documentation of courses limits dissemination and adaptation by other programs. Approaches to reaching a diverse student population via appropriate pedagogy are not documented in curriculum materials, and the effective use of these approaches is unknown. ATE will have to address these deficiencies before the success of the program improvement effort can be fully evaluated.

RECOMMENDATIONS

Based on the data and the conclusions presented above, the following recommendations are provided for NSF-ATE and ATE-funded *projects* engaged in program improvement:

Recommendations for ATE

NSF-ATE program managers should consider the following recommendations (Note: The rationale for the recommendation is presented below each recommendation and is based on the data presented in the paper.):

(1) Continue the increased emphasis on disseminating products that are developed by ATE *projects* and develop a protocol for measuring the impact of these efforts, since materials development is integral to program improvement.

A large number of ATE-supported materials are being developed and implemented by *projects* involved in program improvement. However, the impact of these materials in preparing skilled technicians is small when compared with the size of the U.S. workforce. To achieve a significant impact, more technical education programs need to adapt these materials to their program needs. ATE should consider expanding support of dissemination strategies, including providing guidance on approaches to and funding for clearinghouse-type efforts. *Projects* disseminating materials should be required to provide data on the results of their efforts as a requirement of accepting funding.

(2) Place greater emphasis on pilot and field-testing of the materials developed for the programs by the *projects* and encourage the involvement of business and industry in these activities.

Projects are producing significant quantities of enhanced materials; however, there is minimal verification of these products by the user – business/industry. To ensure materials are meeting their ultimate objective of educating skilled technicians for the workforce, those employing the program completers need to be part of the pilot/field- testing process.

(3) Define minimum standards for documenting materials and require *projects* to meet or exceed these standards.

Based on site visit reports, there is minimal detailed documentation of improved instructional products. This limits the dissemination of these materials to other potential users and constrains the expanded impact of producing more advanced technicians.

(4) Encourage *projects* to identify, document, and use advanced pedagogical approaches to meet the learning needs of a diverse population.

With increased diversity in the U.S. population as reflected in ATE program enrollments, instruction should be provided in ways that meet the varied needs of the population. Data from reports indicate that the pedagogical approaches to ensuring learning in diverse populations are, at a minimum, not documented and may not be given high priority by *projects*.

(5) Fund *projects* that develop replicable strategies for increasing articulation between secondary schools and those associate degree colleges that provide secondary students with advanced college standing.

There is a strong emphasis by *projects* on articulation at the degree level, but limited emphasis on granting advanced standing for secondary students in technical programs. Providing such advanced standing could act to encourage students to continue their studies in advanced technological disciplines.

(6) Define the protocol for routinely gathering program improvement data (including those in underrepresented populations and resulting from product dissemination) and analyze these data in relation to the U.S.'s requirements for skilled technicians.

As discussed, there is no source of verifiable outcomes data for the ATE program. Without these data, the efficacy of the program can be called into question. Although there is often a bias against gathering statistical data, it is essential to "count the beans" if ATE's staff is to answer questions about production and workforce impact.

(7) Develop and fund longitudinal evaluation of the performance of completers of improved programs.

Under the current system of funding, the impact of ATE-educated technicians in the workplace is not, if measured, reported since there is no funding or requirement for longitudinal tracking of program completers. A segment of the ongoing evaluation of the ATE program should be to develop and conduct a longitudinal impact study.

(8) Continue to fund external evaluations of the ATE program, including site visits, to determine an independent measure of the impact of efforts to increase skilled technicians. As part of the evaluation effort, data reporting methodology should be refined and standardized to increase the usefulness of these data.

The current evaluation process of ATE *projects* is producing the only comprehensive data on the impact of the program. As with any complex effort, instruments such as the survey will require refinement. Also, issues about production and impact data need to be addressed through the evaluation effort. Continuation of an independent external evaluation process is prudent.

Recommendations for ATE Projects

The following recommendations are provided for ATE-funded *projects* in program improvement:

(1) Verify the effectiveness of their improved programs.
(2) Involve business and industry throughout the program development and implementation phase of the *project*.
(3) Encourage work between community-college-based *projects* and secondary schools to create advance standing transfer agreements for secondary students satisfactorily completing a block of secondary level instruction in the appropriate discipline.
(4) Create materials and course documentation that can be used by others to replicate or adapt program improvement course/components.
(5) Aggressively encourage institutions with similar human resources development needs to adapt and use materials developed as part of the program improvement project, and mentor this process.

(6) Routinely collect and analyze outcomes data.
(7) Gather and analyze data on the performance of program completers.

STRATEGIES FOR EVALUATION

The items below should be considered for *project* improvement evaluation plans. They also can serve to assist organizations or individuals developing proposals for ATE funding in structuring *project* activities.

(1) Review workforce and associated standards in relation to outcome competencies of the program improvement *project*.
(2) Determine the degree of improvement and effectiveness of STEM instruction.
(3) Evaluate the integration level of STEM, communications, and the development of teamwork and critical thinking skills throughout the curriculum.
(4) Compare equipment used in the program to that used by technicians in industry.
(5) Determine effectiveness of teaching/learning strategies used to meet the needs of a diverse student population.
(6) Gather and analyze program pilot and field-test data.
(7) Gather and analyze outcomes data for all *project* locations.
(8) Determine employer satisfaction with program completers.
(9) Collect and analyze program effectiveness data such as retention and placement.
(10) Review program documentation for adequacy in dissemination, replication, and adaptation efforts.
(11) Create a methodology for tracking completers and their effectiveness as part of a world class workforce.

NOTES

1. *Status Report 1: The Nature of the ATE Program*, p. 1, Kalamazoo, Michigan: The Evaluation Center, Western Michigan University, May 2000.
2. *Ibid.* p. 3, Table 3.
3. The Western Michigan University evaluation project at The Evaluation Center, Kalamazoo, Michigan, has been funded by NSF to conduct an independent evaluation of ATE projects.
4. As a matter of convenience, the term *community college* will be used in this paper when referring to associate-degree-granting colleges.
5. *Advanced Technological Education (ATE), Program Solicitation*. 2000, pp. 5–7.

6. *Advanced Technological Education (ATE), Program Solicitation.* 1997, pp. 4–5 & 2000, p. 7.

7. Articulation as used here is the recognition of educational experience and/or the transfer of "credit" from a lower level to a higher level that results in advanced standing at the higher level.

8. *Advanced Technological Education (ATE), Program Solicitation.* 2000, pp. 11–12.

9. Developmental education efforts are a basic commodity in two-year colleges, which generally have no or limited requirements for enrollment. As a result, this "open door" philosophy requires these colleges to provide remediation of underprepared entering students in order to provide the academic readiness needed for success in the college's programs. This paper will simply note the need to prepare students for success in ATE improved programs, since a discussion of development programs is beyond the scope of the topic of *program improvement.*

10. ATE Site Visit Report Outline, *ATE Drivers*, The Evaluation Center, Western Michigan University. Kalamazoo, Michigan, Distributed September 2000.

11. *Curriculum* as used hereafter in this paper is synonymous with the term "program" in *program improvement.* As used, both curriculum and program are a sequential set of learning experiences, which results in achieving a defined set of specific occupational competencies and, as appropriate, certification or award of an academic degree.

12. Centers are expressly identified in ATE funding and are more comprehensive and funded for a longer duration than projects.

13. In this paper, related academic offerings such as writing, mathematics, science, and speech will be referred to as general education. Appropriate credentialed faculty teach these offerings.

14. The term *technical education* refers to offerings of focused technical content and are generally taught by professionals in the technical content area.

15. *Reverse transfer* is a term used to identify students with a four-year degree returning to a two-year college to gain technical knowledge and skills to increase their employability by business and industry.

16. *ATE Program Guidance*, 1996, p. 12.

17. *ATE Program Solicitation*, 2000, p. 16.

18. The 2000 Survey Report was a final version and the 2001 a draft version. In general, referenced data will be from the 2000 survey. When significant differences between 2000 and 2001 data are evident, this fact will be noted.

19. Site visit reports are comprehensive documents created by members of the WMU evaluation project's visiting teams that did on-site reviews of ATE *projects* at 13 locations. These reports are limited in their distribution to preserve the anonymity of the sites visited. Therefore, no citations will be provided when referring to information contained in these reports.

20. The sources of this review were program descriptions and catalog materials from selected *project* institutions (approximately 35 programs at 15 institutions).The limitations to this review are the limited scope of the review, the lack of a "before improvement" benchmark to measure change, and the brevity of course descriptions. In spite of these limitations, useful information on STEM integration in program improvement was gained. To preclude identification of specific projects, direct references will not be provided.

21. The reported data on levels of program improvement can be misleading. Although 93% of the respondents were involved at the associate degree level, only half were

exclusively at that level. The remaining 40% were at multiple levels of secondary-associate or associate-baccalaureate levels.

22. An issue requiring more data and analysis is the degree students are required to and do follow the program sequence.

23. ATE guidance to institutions seeking funding states "that all programs have a core of courses in science and mathematics to serve as prerequisites or co-requisites for specialized technology courses." *ATE Program Solicitation*, 1998, p. 3.

24. DACUM (Developing A CUrriculuM) is a structured way of identifying worker duties and tasks via a group process using "expert workers" from industry. The process is often used in community colleges to determine standards for both new and revised program development. Some effort has been made to adapt DACUMS to predicting future technical education needs, but the lack of expert workers makes such use marginally effective.

25. As used here "readiness" is the level of mathematics proficiency a student must demonstrate by either testing out of or passing an appropriate bridging course prior to enrolling in the specified curriculum course.

26. Technical mathematics is a course designed for career programs that cover topics needed in technical applications. Topics range from arithmetic functions such as fractions and decimals to basic algebra, trigonometry, geometry, and statistics. These courses are normally not transferable to four-year programs.

27. It should be noted that only one engineering technology program sequence was reviewed by site visits and in the document review. The mathematics literacy of technicians in these fields could be significantly higher than those discussed above.

28. This holds true for both credit and noncredit continuing education offerings.

29. Students who already possess a college degree or have had substantial college-level general education.

30. *ATE Program Solicitation*, 2000, p. 8.

31. *ATE Program Solicitation*, 2000, p. 17.

32. *Underrepresented populations* as used here refer to the proportional representation of categories of workers in a particular occupation in relation to the general population. This includes race, ethnicity, and gender.

33. The total percentages of minorities plus whites would be expected to equal 100%. Obviously, in reporting data, *projects* did not ensure their data did so; thus, the totals do not reach 100%. Additionally, in future surveys a benchmark of "prior to ATE" would be useful in judging change in diversity enrollments.

34. The distributions reported may be historical for the institutions, may reflect the mix of the area's population or, since they are an aggregate, may range widely for each reporting institution.

35. *ATE Program Solicitation*, 2000, p. 8.

36. *Comparison of the 2000 and 2001 Survey Findings*, The Evaluation Center, June 2001, p. 9. Draft excerpt from *Survey 2001: The Status of ATE Projects and Centers*.

37. A more sound statistical approach would have been to exclude the reported student populations that fell significantly outside the nominal range from the reported averages. Another approach might have been to include the median enrollments as well as the average to reveal the variation between the two measures of central tendency.

38. It should be noted that these data are based on a single program per project selected by the project leaders and are not a structured or controlled sample. As a result, the reported averages may or may not be typical of all programs.

39. Workforce vs. education are not mutually exclusive categories, and students can be in both categories. Hence, the total can exceed 100%.

40. Based on current Bureau of Labor Statistics (http://stats.bls.gov/), there are 33 million employees in the U.S. workforce with some college, but less than a bachelor's degree (10.7 million possess an associates degree). Although no statistical data identifies "technicians," it appears safe to assume that a significant number of the 33 million are in skilled advanced technology jobs. If this assumption is accurate, then the challenge of upgrading these technicians and producing more cutting-edge technicians is an immense undertaking.

41. *ATE Year 2000 Program Solicitation*, pp. 10 & 13.

42. Current data gathering by the WMU evaluation project is not structured to obtain global statistics or impact information for materials disseminated by projects and centers. Revised protocols would be required to obtain these data.

43. Transfer of learning experience (versus academic credits), although not widespread, does afford the potential for more articulated transfers. However, academic credit remains the "coin of the realm" for the vast majority of degrees.

44. Historically, community college students have been able to transfer their lower division (freshman and sophomore level) undergraduate general education credits to baccalaureate programs. Most states mandate such transfer for publicly funded state institutions. On the other hand, technical credit transfer is normally an institution-to-institution negotiated agreement for individual courses (and sometimes for individual students). Even between same level institutions, general education credit transfer is more common than transfer of technical credit. The issue is generally the commonality of technical content. Advanced standing in associate degree programs based on secondary coursework does occur based on individual institution to institution agreements but is not, in the author's experience, widespread.

ASSESSING PROFESSIONAL DEVELOPMENT

Norman Gold and Karen Powe

EXAMINING ISSUES IN PROFESSIONAL DEVELOPMENT: AN OVERVIEW

As secondary and postsecondary students think about careers that may interest them and as educators plan the curricula to prepare their students for the workplace, the continuing evolution of technology affects both. The tech sector includes a wide array of science- and math-based occupations with all five computer-based occupations projected to be the fastest growing fields in the first decade of the 21st century: computer engineer, computer support specialist, computer systems analyst, database administrator, and desktop publishing specialist (*Occupational Outlook Quarterly*, 2000, Summer). Much of the burden for producing the new generation of technology professionals has fallen on community colleges and their faculties, which are being asked to provide analogous training to students, industry employees, and university instructors. To help meet the rapidly expanding demands for technology training, the Advanced Technological Education (ATE) program has targeted significant funding for professional development of community college faculties.

Current ATE programs are an essential element of this paper, which is intended to provide information and guidance for community colleges developing

Evaluating the Upgrading of Technical Courses at Two-Year Colleges:
NSF'S Advanced Technological Education Program
Advances in Program Evaluation, Volume 9, 105–130
Published by Elsevier Ltd.
ISSN: 1474-7863/doi:10.1016/S1474-7863(04)09006-4

and/or expanding their technology education programs. The paper will take a three-pronged approach:

(1) Examine what is happening in professional development programs and what more is needed.
(2) Review professional development literature for the state-of-the-art[1] on training on the use of educational technology in the classroom.
(3) Evaluate the impact and effectiveness of the ATE professional development programs.

Experts in the fields of technology education and professional development who have reviewed the current situation broadly agree on the needs and a set of responses. To meet the demands of students for a first-class education and of employers for highly skilled workers, community college faculties must keep pace with rapidly changing technologies. High quality, ongoing professional development for faculty is imperative. Yet, according to Tenbusch (1998), "National statistics have shown that [instructors] receive far less on-the-job training in technology than any other group of professionals" (*Electronic School*, 1–6). The U.S. Congress, Office of Technology Assessment (as cited in Brand, 1997), reports it is the lack of staff development that is the primary stumbling block in providing excellent technology instruction.

If our primary educational goal is to prepare students for a technological world and, if we believe that instructors are the first learners, then providing faculty with state-of-the-art professional development becomes a top educational priority (Hord, 1997). It is no longer acceptable to present an occasional staff development workshop. Indeed, professional development should be part of the daily life of instructors with time provided to work in collaborative groups, conduct research, give and receive mentoring, and enhance knowledge and skills. The model of staff development for technology must put the instructor/learner at the center of the learning experience and provide a meaningful context for learning (Stager, 1995).

It is the instructor/learner and their institutions that are the primary concern of this paper. In the following section, we review current literature about the state-of-the-art on training on the use of educational technology in the classroom, looking at three issues:

(1) What is happening and what is needed in professional development at the community college level.
(2) What we can learn about professional development from industry and service provider trainers.
(3) What makes up exemplary professional development programs.

SURVEYING THE SCENE: FACTORS THAT LIMIT PROFESSIONAL DEVELOPMENT

Perhaps the biggest problem in the area of professional development for educators, particularly at the postsecondary level, is the very lack of it. The scarcity of staff development tends to result from three factors that emerge over and over in the professional development literature.

(1) *Time, or more accurately stated, the lack of time is one of the most difficult problems faced by schools* (Watts & Castle, 1993). In fact, Fulton and Miles (as cited in NCREL, 1997) say that time is the key issue in every analysis of change in education in recent years. Cook (1997) adds: "A fundamental lesson learned in the past decade of school reform efforts is that far more time is required for professional development and cooperative work [among faculty members] than is now available." Corcoran (1995, cited in NCREL, 1997) stresses that faculties "need more time to work with colleagues . . . and to revise curriculum."

(2) *Both the public and policymakers fail to give professional development a high priority*. It is a commonly held belief that faculty members should know what they need to know before they begin to teach and should, therefore, spend their time in the classroom. "[T]he public and policymakers perceive [instructors'] work has not changed. They continue to think [instructors] are working only when they are with their students . . . Education must respond to the changing needs of students and [faculty], just as business has reacted to its changing needs by implementing employee training" (Darling-Hammond, 1991).

(3) *Professional development is lacking not only in quantity, but also quality*. Community colleges have tended to undervalue professional development, making faculty members responsible for their own continuing education. In reality, if staff development for instructors is to be truly effective, administrators must not simply pay lip service to the cause. They must take supporting action (Persky, 1990).

It is clear that the traditional professional development model for educators is inadequate to meet the existing need for technology training among community college faculty. Traditional professional development, according to McKenzie (1991), typically has several all-too-well-known components:

- One-size-fits-all training.
- Occasional half-day or day-long workshops.
- Limited time to practice new skills/employ new knowledge.
- Little or no follow-up to workshops.

- Lectures from outside experts about the latest educational trends.
- Little or no compensation for time spent.
- Ineffective planning and inadequate budgets to provide training that will transfer to the classroom.

Jamieson McKenzie, editor of *The Educational Technology Journal*, states that making the change from traditional staff training is the fundamental issue in providing effective professional development: "Shifting from industrial age thinking and teaching to information age thinking and teaching is as dramatic an adjustment as shifting from teaching in a classroom to teaching underwater. The training agenda is no simple list of skills; everybody must learn an entirely new approach" (April 1991).

In regard to poor planning, McKenzie says: "... training often occurs at the wrong time of the day in a room that is either too hot or too cold, and the instructors are often expected to subsidize the learning process with their own time and money ... Quite a contrast with training in private industry that takes place in comfortable training centers or hotels with good food and superb session leaders!"

Professional development literature has consistently stressed the "continuing" part of continuing education. Since the 1980s, staff training researchers and practitioners have talked about ongoing training. They have urged follow-up sessions in the weeks and months after workshops to sustain new practices. They have preached the importance of encouraging and supporting instructors (through in-classroom coaching by trainers and peer leaders) to continue learning and implementing new knowledge and skills, lest the momentum for change be lost (Bents & Howey, 1981; Joyce & Showers, 1983; Sprinthall & Sprinthall, 1980).

When follow-up has occurred, half the participants in ATE programs report trying training materials in their classrooms and one third report implementing them (Gullickson et al., 2000). These responses suggest that more follow-up is required to support instructors in trial and implementation of new materials and teaching techniques. Indeed, the need for concentrated attention to professional development training and follow-up is picturesquely described by McKenzie: "Greater time and resources must be devoted to teacher learning, and greater attention must be given to the needs of teachers as adult learners. A generation of teachers who view themselves as pioneers, inventors, and discoverers must be nurtured so that when the waves of the future hit the shores of our present our teachers will dive headlong through them rather than ducking, running for shore, or allowing themselves to be swept away."

Over the past decade, much has been learned about what makes up effective professional development and that overcomes the deficiencies listed above. The next section identifies what has been learned about the aspects associated with more effective professional development.

IDENTIFYING QUALITIES THAT DEFINE EFFECTIVE PROFESSIONAL DEVELOPMENT AND DESIGNING AN ASSESSMENT STRUCTURE

The need for concentrated attention on crafting and instituting effective professional development programs is perhaps best stated by Sparks and Hirsh (1999): "We cannot expect [instructors] to teach what they do not know, nor to use yesterday's training to prepare today's students for tomorrow's future. We certainly cannot expect our [instructors] to share and learn from each other's knowledge and skill unless we provide them with the research, structures, time, and money with which to do it. Ultimately, quality staff development benefits students by channeling the talents and expertise of all the school's faculty in all the school's classrooms. By improving staff development . . ., "we will be helping all [instructors] to excel at helping all students reach the high levels of achievement they need to succeed."

Current professional development literature delineates the basic qualities that comprise effective and successful staff training. Consistently, seven elements emerge as necessary for high-quality professional development programs:

- Ongoing learning and training.
- Institutional support.
- Hands-on and classroom-based experiences.
- Individualized training.
- Follow-up training.
- Mentoring.
- Train-the-trainers approaches to continuing education.

Here is what the literature tells us about why these elements are important:

Element 1: Ongoing Learning and Training

Stager (1995) and Dobbs (2000) stress that professional development should be a part of instructors' daily lives. Instructors must have the time to gain new knowledge and skills for courses in a wide range of technology fields with training and development on a continuing basis to keep up with changes. Staff development for technology requires adaptable training content and sufficient opportunity for working with the tools over time (Guskey, 1999; Sparks & Hirsh, 1999).

Corporate America has long recognized the value of ongoing learning, and companies are promoting it among their employees (Dobbs, 2000, January). Dean Spitzer, senior performance consultant with IBM, suggests staff development strategies for education:

> [T]o affect [instructors'] content knowledge, instruction skills, and student learning . . . the staff development effort [must be] sufficiently powerful to accomplish those purposes. [E]fforts must be sustained over months and years, provide a great deal of in-classroom demonstration and coaching, and offer generous amounts of time for small groups of faculty members" to work together developing curricula, instituting and evaluating new instructional methods, and solving common problems. "Nothing less will get the job done" (cited in Sparks, 2000, March).

Guskey (1986) and more recently Kimmel et al. (1999) note that it has become accepted that long-term intensive professional development programs are necessary and that short in-service programs or workshops are not sufficient to produce sustained change (Guskey, 1986). Conferences, workshops, and in-service training, however, continue as the most popular forms of professional development in the ATE program. Assessment of the current ATE sites indicates that 15% or less provide other learning activities, internships, or online courses (Gullickson et al., 2000).

Element 2: Institutional Support

Persky (1990) notes that administrative support must be strong to ensure effective professional development. Despite resource constraints, time and money is required for good staff training. Tenbusch (1998) acknowledges the concerns of educators and administrators regarding resources for professional development but, he says, it's possible. In an article written for Electronic School magazine, Tenbusch notes: "The business community knows that for every dollar spent on hardware and software, another dollar must go toward staff development. Developing a successful [instructor] technology-training program requires more than turning [the] faculty loose after a few workshops. . . . [A]dequate training for [instructors] can be expensive, in terms of both time and money. But if [schools and community colleges] don't do a better job of allocating resources for professional development – instead of putting all the budget into technology acquisition – [they] will be left with the tools but not the talent to prepare [students] for a technological world."

Faculty support is also a piece of the training pie, but it is often missing. Tenbusch (1998) offers some field-tested strategies for building faculty support for and interest in technology training: A successful professional development program in technology must focus on enhancing knowledge and giving faculty members the incentives to expend the time and energy to learn what they want and need to know. There are four basic parts to an effective training program: (1) intensive training with opportunities to work with new ideas and materials over several sessions; (2) follow-up with trainers and/or mentors over an extended time period; (3) time to consult with colleagues; and (4) occasional observation of other instructors who use exemplary techniques.

Effective staff training, then, requires that administrators and instructors establish the structure that will allow professional development to occur. That structure includes:

- Setting schedules that allow for flexibility.
- Providing opportunities for instructors to complete the development sessions on their own time, at their own pace.
- Offering a combination of learning circumstances, such as traditional workshops and in-class collaboration.
- Creating sessions around small groups for more individualized attention.
- Using instructional variety to present information and teaching skills.

(Browne & Ritchie, 1991; Harvey & Purnell, 1995; Stager, 1995).

Element 3: Hands-on and Classroom-Based Experiences

State-of-the-art staff training is hands-on, classroom-based, and student-centered. It presents analytical problems, using inquiry techniques; it promotes modeling; and it relies on instructors and development professionals working together to create plans, present information and skills, evaluate and redefine education programs, and present refined curricula to meet student and marketplace needs. And, it requires substantial time for instructors to acquire and, in turn, transfer new knowledge and skills to their students (Brand, 1997; Guhlin, 1996; Shelton & Jones, 1996).

Element 4: Individualized Training

Because instructors vary in their levels of expertise at the time of their training, the context, which surrounds their technological professional development, must provide a non-threatening environment that is sensitive to the individual instructor's level of expertise and experiences (Browne & Ritchie, 1991; Shelton & Jones, 1996). Instructors must have significant blocks of time in order to acquire and apply the knowledge and skills necessary to effectively use and teach technology. They also need time for collaborative learning and time to develop the networks that promote collegial learning. To help faculty members complete the "learning cycle" of computer-related professional development, training must be ongoing and systematic (Brand, 1997).

The economics and logistics of staff development will continue to push us to train instructors in groups, but we must remember that the technology itself allows us to individualize instructor learning and provide support in ways that offer new economies of scale. For example, colleges can provide Internet courses that faculty members can take on their own, videos for at-home instruction, one-on-one on-line mentoring, and other forms of personalized, just-in-time, just-what's-needed formal and informal instruction (Fulton interview by Sparks, 1998).

Element 5: Follow-up Training

A part of ongoing training is the follow-up to professional development opportunities and activities. Being able to work with mentors and colleagues to reinforce new knowledge, skills, and techniques augments both what has been learned and the confidence to use it. There is considerable agreement that collaborative group work and learning is the most powerful kind of reinforcement in professional development (Arter, 2001; Garmston, 1999; Johnson & Johnson, 1999). In several studies, instructors cite the opportunity to collaborate as the most important factor in instituting change. Research evidence also indicates that learning in groups significantly improves learning for adults as well as for children (Arter, 2001).

As a result, Browne and Ritchie (1991), Persky (1990), and Stager (1995) suggest that collaborative problem-solving and cooperative learning must undergird the approach to technology learning for instructors. Although a number of approaches are available, peer coaching and modeling have been most effective in transforming workshop information to classroom applications and practice (Browne & Ritchie, 1991; Persky, 1990).

Element 6: Mentoring

Perhaps the greatest challenge for instructors is putting newly gained knowledge and skills into action. Doing so is greatly enhanced by having someone to guide and advise, someone with whom instructors can discuss and plan – a mentor. Unfortunately, providing mentors is not a part of most professional development programs. The literature suggests that the lack of emphasis on mentoring results from misconceptions about professional development for educators. Public assumption (and sometimes the assumption of administrators) that instructors are only working when they are with their students is a barrier to establishing staff training programs and to providing the support systems necessary for training success. International studies that have compared the daily activities of instructors in other countries have shown that staff development has considerably greater importance abroad. For example, in Japan instructors teach fewer classes than their American colleagues and they use the added time to plan, meet with peers, work with students one-on-one, and participate in professional development (Darling-Hammond, 1994, November).

Susan Loucks-Horsley of the National Research Council agreed, pointing out that pre-service training, professional development, and ongoing support for instructor learning and development are often more robust in other countries. "In many places, [instructors] have fewer student contact hours and more time to work together. While it would be a mistake to try to pattern ourselves after other countries because of the vast cultural and population differences, exploring what we can learn from the rich data base can help us look at ourselves and view images of what can be different" (Sparks, 1997).

The need for concentrated attention on crafting and instituting effective professional development programs and for sharing best practices is, perhaps, best stated by Sparks and Hirsh (2000):

"We cannot expect [instructors] to teach what they do not know, nor to use yesterday's training to prepare today's students for tomorrow's future. We certainly cannot expect our [instructors] to share and learn from each other's knowledge and skills unless we provide them with the research, structures, time, and money with which to do it."

Element 7: Train-the-Trainers

A train-the-trainers approach to continuing education can provide more instructors with more opportunities to learn and be a means for conserving limited professional development resources. In addition, business offers an additional incentive: the observation that colleagues provide each other with a great deal of "informal learning" that may not occur in staff training programs. The Education Development Center, Inc. (EDC) conducted a 2-year study of U.S. corporate cultures, including workers' training. One of the findings of the study was that 70% of what employees know about their jobs, they have learned informally from the people with whom they work. The lesson: Education institutions must provide opportunities for instructors to spend time with colleagues so that they can learn from them and their experiences (Dobbs, 2000). Indeed, colleges should encourage and support the development of interactive faculty and provide opportunities for training to occur formally and informally. In the words of a recent report funded by the Benton Foundation ("The Learning Connection: Schools in the Information Age"): "[W]e must build a human infrastructure at the same pace we are building computers and wiring" (Sparks, 2000, an interview with K. Fulton).

INTEGRATING THE ELEMENTS: EXAMPLES OF BEST PRACTICES

While many programs exist to help elementary/secondary teachers infuse technology into their curricula, professional development for postsecondary technology instructors is largely absent or incomplete. A review of the American Association of Community Colleges' library and online resources indicates that comprehensive professional development programs at community colleges are indeed few and far between.

On the other hand, two programs that receive frequent praise are early ATE participants. The first is Northwest Indian College in Bellingham, Washington. The college developed a team teaching approach with Western Washington University and conducted faculty-training programs stressing the fundamentals of coordinated studies and learning communities. The positive effects on student learning convinced instructors and administrators that, while the cohort approach to teaching and learning is very time-intensive, the program should be retained. Faculty, administrators, and students reported more integrated curriculum, student-centered teaching techniques, and hands-on learning – all credited for increased student commitment, higher retention of students and staff, and more students completing their studies (Mahoney & Barnett, Eds, 2000).

The second project is a three-phase program to develop improved technology programs and teaching methods at Queensborough Community College (QCC) in New York City. Using telecommunications technologies as a means for instructional change, QCC has focused significant effort on faculty enhancement. The goals are to provide science and technology students with more marketable skills, develop methods to keep curricula on the cutting edge, and to improve faculty skills.

QCC's project has been based on the belief that traditional approaches to improvements in instruction have simply attempted to put new technology on top of existing teaching techniques. Doing so has not proven very effective in providing students with the knowledge and skills they need, and the college recognized a pending critical shortage of successful science and engineering technology students unless instructional methods became more effective. The resulting professional development at QCC has served not only the college's own faculty, but has drawn participants from across the United States, Puerto Rico, Japan, and the Netherlands to audit and observe staff training. The college has experienced extraordinary success with technology integration and student retention/completion (Mahoney & Barnett, Eds, 2000).

There are many other technology experts, researchers, and trainers who, from their experience and expertise, can provide diagrams for model professional development. One such group, the Southern Technology Council, reviews the best of professional development programs and notes best practices. Reiterating advice from business, the Council notes, "Mandates and incentives are dominant themes in establishing successful professional development practices." Brand (1997) points out that the need to allot time for continual learning is echoed in studies outside education, which suggest that providing workers with high technology ultimately fails if employees do not receive adequate training and continual, on-the-job support. Other best practices include those listed below:

- Requiring instructors to earn in-service credits in continuing technology training.
- Providing monetary incentives for professional development in technology.
- Offering stipends to faculty who conduct workshops on a piece of software or an application with which they are expert.
- Rewarding faculty members who receive training with free hardware or software.
- Offering interest-free financing to instructors who want to purchase/upgrade personal computers.
- Allowing instructors to take school laptop computers off campus.
- Requiring faculty members who are interested in receiving free Internet accounts to attend training (Southern Technology Council, cited in Tenbusch, 1998).

Several corporations and business associations recognize and practice such supporting techniques with their employees as part of an effort to motivate workers to take part in professional development. Studies in the business sector indicate that providing workers with highly developed technology training fails if the employees do not receive adequate incentives (Brand, 1997). Putting their money where their mouths are, several corporations and business associations throughout the U.S. are working with secondary and postsecondary institutions to share what they have learned. They hope to increase professional development in technology and to promote the transition of students from education to work. Organizations with active partnerships include:

- American Airlines.
- Consortium of Allied-Signal, Goodyear, Honeywell, Hughes, McDonnell-Douglas, and Motorola.

- Institute for Women in Trades, Technology, and Science.
- National Association of Automotive Dealers.
- National Association of Manufacturers.
- National Center on Education and the Economy (http://proquest.uni.com/pqdweb).

Some bottom-line advice comes from IBM consultant Dean Spitzer. He suggests that professional development needs to be based on the driving concerns of the business – revenue, profits, customer satisfaction, market share, cycle time, production quality, and employee retention rates (Spitzer, 1999, June). For education, that means looking at staff development in terms of its intended outcomes:

- Faculty satisfaction with what they have learned.
- Student satisfaction with what they learn.
- Student retention.
- Student referrals that bring in other students.
- Professional development time spent as compared with the results.
- Faculty retention rates.
- The community college's market share of students.

While the role of community colleges is to provide educational opportunities and prepare students for the future, the schools have to be monetarily successful in order to maintain their programs.

The American Association of Community Colleges (AACC) points out that constraints on public funds show no signs of abating, as more legislatures predicate budget increases upon performance of specific goals. Competition is pushing community colleges to consider students as customers. In turn, emphasis on students' learning is growing as employers and society expect students to demonstrate competence in what they have been taught. Assuring the knowledge and expertise of community college faculties must be a critical part of every successful school's plan for the future (AACC Research, 2000).

ASSESSING PROFESSIONAL DEVELOPMENT ACROSS THE ATE SITES

Ideally, every professional development program will be based on research and needs assessments and will incorporate the basic elements that make up effective training programs. Those elements, described in the "Identifying Qualities That Define Effective Professional Development" section, are listed below:

(1) Ongoing learning and training.
(2) Institutional support.
(3) Hands-on and classroom-based experiences.
(4) Individualized training.
(5) Follow-up training.
(6) Mentoring.
(7) Train-the-trainers approaches to continuing education.

Assessment of the 13 sites visited by the WMU evaluation project reveals a varying adoption of the 7 elements of effective professional development. Oftentimes, the institutions have introduced 1 or 2 elements of a training program, planning to add others as they progress. Practically speaking, organizations may have to decide which professional development elements are more important to them and which their resources will permit them to introduce. It is essential, however, to keep in mind that the fewer elements present in the training program, the less likely it will be that desired results will occur and/or that the program can be sustained.[2]

Element 1: Ongoing Learning and Training

Professional development should be ongoing, a part of instructors' daily lives with time to enhance their technology knowledge and skills, to learn new methodologies, and to work in collaborative groups (Dobbs, 2000; Stager, 1995). Twelve of the 13 ATE sites report ongoing professional development.

ATE Site 1 has identified the need for instructors to receive initial and ongoing training as essential to effective program implementation and dissemination.

ATE Site 2 recognizes the need for continued professional development. The only questions are what form that development should take and how it will be funded. To date, college funds have not been sufficient to provide appropriate faculty training. ATE funding has made access to professional development possible and future funding is assured from the site's county.

ATE Site 3 has focused on ongoing professional development opportunities. Instructors have continuing workshops and training experiences throughout the year, presented by knowledgeable trainers in content areas that have been identified by the participants as needed. Site 3's approach assures current classrooms by keeping faculty members up-to-date.

ATE Site 4 has begun a professional development training and support program for faculty members who want to change their instructional methodologies. The program plans continuing assistance as an integral part of the staff training. A central element of the ongoing training is the planning for new criteria to be applied as instructors are added and/or replaced.

ATE Site 5 has established a 3-year teacher-training program that includes workshops modeling curriculum and laboratory environmental design components. Thirty instructors attended an inaugural 4-week workshop and formed the core of instructors for future workshops. Workshop instructors must teach network curriculum at public, 2-year colleges and are required to earn industrial network management certification. In order to receive a full stipend, workshop participants must pass industry examinations for certification.

ATE Site 6 has designed a model professional development workshop. Attended by high school and community college physics and science teachers and instructors, the workshop includes teaching techniques, innovations in materials, business needs and perspectives, and on-site experiences with companies in the technology area covered by the site. In addition to the camps, Site 6 is planning faculty internships and expanded funding for additional professional development training and travel.

ATE Site 7's ongoing professional development includes summer institutes and continuing opportunities for instructors to share ideas and to work with their peers and local industry representatives. Instructors report that the ongoing activities have enabled them to gain supplementary resources throughout the life of the *project*.

ATE Site 8 has provided technology training for full-time and some adjunct faculty members and has ensured that instructors are kept up-to-date through regular professional development training each year.

ATE Site 9 conducts summer professional development institutes and facilitates numerous meeting and activities designed to effect systemic changes in technological training program at secondary and postsecondary levels.

ATE Site 10 provides commercial courses and a series of workshops on curriculum, assessment tools, pedagogy, and networking. Instructors note their approval of the high quality of the professional development.

ATE Site 11 has ongoing training that is designed to emulate workplace situations and illustrate the principles of effective use/management of technology. Development of materials is continuing, and several sites serve as pilots for field-testing the materials.

ATE Site 13 has held more than 40 workshops in 7 regions, focusing on workplace experiences and expanding instructors' technology expertise. In addition, summer forums keep instructors up-to-date with biotechnology programs.

Element 2: Institutional Support

Institutional support is crucial to providing professional development. Current literature reports and some project sites confirm that community colleges have tended to undervalue faculty training. If staff development is to be effective, administrators must take supporting action (Persky, 1990). Nine of the 13 ATE projects report strong institutional support.

At ATE Site 2, professional development is recognized as a high need by college administrators, but limited by available funding. The college is working with its county to assure future funding for ongoing staff training.

ATE Site 4 administrators have recognized that broad support is necessary to provide effective development opportunities. Having adopted a special program as an innovative concept of teaching and learning, the university has involved its division related to this program and distributed the workload associated with developing an innovative program. As a result, the division and the college both benefit. The institution is committed to implementing similar staff development programs throughout the university.

ATE Site 5 has initiated a professional development program that is ongoing, individualized, and hands-on, establishing an infrastructure to train instructors. Strong institutional support has enabled Site 5 to engage 110 public schools and 23 community colleges in workshops, impacting 67% of the state's 82 counties.

ATE Site 6 has worked closely with industry to establish an education-business alliance to address technology needs. The consortium's objectives are to respond in a coordinated fashion to the rapidly evolving industry requirements for a trained high-tech workforce and to share the limited educational resources to develop new training materials and laboratories.

ATE Site 7 has been encouraged by an industry needing technicians. Providing effective professional development is a stated goal of the college administrators and is galvanized by strong public support for its programs.

ATE Site 8 has provided funding for technical training of full-time and some adjunct faculty. While most colleges have limited professional development funds, Site 8 has worked to overcome the budget barrier by sending a limited number of faculty members to attend training and, then, to report to other instructors at monthly faculty meetings. All campuses have faculty who have certification training. Each Site 8 campus reinforces the importance of professional development by requiring a minimum number of hours each year, partly basing salary raises on staff training completed, and providing pedagogy and course/curriculum design workshops.

ATE Site 9 has supported professional development by making funding and time available to instructors to attend national meetings and workshops designed to help them develop their content expertise.

ATE Site 10 recognizes the need to provide professional development for its instructors. This support is reflected in the *project's* statement of intent: "to address the critical need for a well-trained workforce with the diverse skills needed in 'technology.'" While faculty members note the need for more release time for learning and working with complicated technology, they commend the training courses.

ATE Site 11 faculty praises the institutional support for the professional development they have received. Instructors rate their training as organized, effective, and useful.

Element 3: Hands-on and Classroom-Based Experiences

State-of-the-art professional development is hands-on and classroom-based. It requires the training and time for instructors to acquire, gain confidence with and, in turn, transfer new knowledge and skills to their students (Brand, 1997; Guhlin, 1996; Shelton & Jones, 1996). Eight of the 13 ATE projects report hands-on, classroom-based professional development.

ATE Site 1, believing that both instructors and their students will be more actively engaged by learning in an applied setting, has made classroom-based technical assistance the center piece of its professional development efforts.

ATE Site 3's professional development component emphasizes hands-on module user training and maximum exposure to new information and technology. In addition, faculty members are encouraged to "hear what is on the street" by attending the professional conferences of the industry.

ATE Site 4 summer internships for faculty are under discussion by local industry and the university. Faculty members have expressed interest and the program's industrial advisory committee has said it could support faculty internships. Among the issues to be resolved is scheduling so internships will not interfere with current teaching calendars.

ATE Site 5 has access to a dedicated laboratory to support the hands-on portion of staff training. Faculty members are increasingly taking advantage of the facility and the industry-training program for its curriculum. Site 5 has eleven 2-year college education and training sites and 2 university education and training sites. Continued qualification training for 2-year faculty is available through the programs institutionalized by the colleges, including an academy, online courses, and short courses.

ATE Site 6 has designed a professional development program that includes on-site experience in industry production and lab facilities, interaction with industry representatives, and participation in internships.

ATE Site 7's professional development includes training institutes and follow-up meetings, but has also provided much hands-on experience or classroom-based assistance. While the institute exposed instructors to relevant technology content, it did so through tours and field trips to relevant technology facilities. Several instructors have expressed an interest in having professional development aimed at pedagogy, curricular materials they could use in their classrooms, and/or time to develop materials collaboratively.

ATE Site 12 has provided workplace experience for instructors through academies, as well as ongoing access to technological expertise and resources.

ATE Site 13 has focused on providing workplace experience and strengthening the technological know-how of its instructors. Concentration on the need for constant upgrading in the technology focus of the site has provided faculty members with exposure to and experience with current technology.

Element 4: Individualized Training

Individualized training ensures that instructors can learn new technologies and new instructional methodologies that are consistent with their current levels of expertise. The one-size-fits-all approach to training is not effective in education technology training (Browne & Ritchie, 1991; Shelton & Jones, 1996). Six of the 13 ATE sites report individualized training.

ATE Site 2 has provided professional development via workshops or credit or continuing courses. Faculty members have expressed a high level of satisfaction with the courses and have used the knowledge/skills gained in developing several technical courses. Given the diversity of the technical topics in the program, this individualized approach seems sound.

ATE Site 3 offers training in content areas that have been identified by instructors and are presented by trainers who are knowledgeable and current in the industry. To augment the introduction of new information and materials, previously trained colleagues share their ideas for using new materials in the classroom. The faculty-to-faculty workshops reinforce the ongoing nature of the Site's professional development.

ATE Site 5 has a self-paced professional development program that is designed to allow optimal learning and ensure the learner's confidence with new information and technologies before s/he attempts to teach the subject matter.

ATE Site 6 has designed a workshop specific to instructors' needs. Workshop activities include the industry perspective, aspects of the science involved in the

Site's technology area, on-site experience with production and lab facilities, and discussion of techniques for integrating the technology area into classroom instruction.

ATE Site 9 conducts summer institutes that focus on instructor needs in terms of implementing new ideas and materials into their curricula.

ATE Site 12 has adopted 5 goals, one of which directly addresses individualized training: to provide workplace experience for instructors, along with ongoing access to technological expertise and resources, for the purpose of enhancing instructor knowledge and skills and ensuring classroom learning.

Element 5: Follow-up Training

Follow-up to training assures instructors that continuing advice and assistance are available as they implement new systems and techniques. Instructors must know they have support in adapting training content and applying new tools (Guskey, 1999; Sparks & Hirsh, 1999). All 13 ATE *projects* report follow-up activities.

ATE Site 1 is developing instructors' guides and workshops intended to expand opportunities of faculty members to access technical assistance on a continuing basis.

ATE Site 2 has established, in addition to continuing courses in the Site's technology area, a business/industry advisory board that assists faculty in defining changing technology competencies and developing new curricula.

ATE Site 3 has provided a strong communications network to support instructors and offers technical workshops throughout the academic year. In addition, faculty members are encouraged to attend national conferences, valued for their training opportunities and networking experiences.

ATE Site 4's approach to follow-up is 3-fold: to provide ongoing training for faculty members who want to change their instructional methods, to develop new criteria for teaching technological programs as instructors are added/replaced, and to make summer internships available for faculty members.

ATE Site 5 has been instrumental in creating a web-based program of study in its technology area. The target audience includes faculty pursuing skills and certification as NT instructors. The instructor component is available free-of-charge as part of Site 5's continuing professional development efforts.

ATE Site 6 has built an education-business alliance that ensures appropriated follow-up training activities by coordinating responses to evolving industry requirements and sharing limited resources to develop new course materials.

ATE Site 7 holds follow-up meetings devoted to developing articulation agreements that will ensure collaboration between secondary and postsecondary levels and promoting continuing professional development for high school teachers and community college instructors.

ATE Site 8 ensures that follow-up will occur through their faculty members who have certification in its technology area. These instructors are able to provide up-to-date technological instruction for their colleagues. In addition, workshops in pedagogy and curriculum design are provided through the Site's development program.

ATE Site 9 facilitates meetings and activities throughout the academic year that bring together various individuals and organizations. These events serve as a catalyst for professional development experiences for instructors.

ATE Site 10 encourages interaction between instructors by e-mail. Faculty members report that they share ideas and support each other on an ongoing basis.

ATE Site 11 provides continuing workshops for instructors to work on activities, modify teaching techniques, and hear how the *project's* technology area is used in business/industry settings. Suggestions for instructor participants in these follow-up sessions have resulted in production of a video of a business/industry scenario and annotated exercises to accompany instructional materials.

ATE Site 12 has a working education-business partnership that is committed to follow-up. The Site's intention is stated in its third program goal: to provide workplace experiences for instructors, along with ongoing access to technological expertise and resources for the purpose of enhancing learning.

ATE Site 13 acknowledges the importance of follow-up training and has set 4 objectives to direct its efforts: (1) continuing improvement of faculty technical skills, scientific knowledge, and pedagogical approaches; (2) bringing workplace experiences into the classroom; (3) developing selected courses and instructional materials; and (4) promoting and improving evaluation of program qualities and effectiveness.

Element 6: Mentoring

Mentoring exposes faculty members to the experience of colleagues, provides a safe place to try out new ideas, and significantly increases development of creative instructional strategies (Stager, 1995; Tenbusch, 1998). Only one ATE *project* reports mentoring programs.

ATE Site 7 has included a summer institute and follow-up meetings in its professional development to explain the mentoring program, which includes a new externship for instructors.

Element 7: Train-the-Trainers

A train-the-trainers approach can expand training opportunities and alleviate the constant time and money problems faced by schools and colleges. In addition, faculty members who can act as trainers bring an important factor to professional development: They can identify their colleagues' current interests and needs and provide training that is specifically geared to address those interests/needs (Brand, 1997; Shelton & Jones, 1996). Four of the 13 ATE sites report using the train-the-trainers approach to expand professional development opportunities.

ATE Site 1 encourages instructors who have received technology training to train their colleagues, in turn, thereby expanding programs in their colleges. To make training more reliable and efficient, a train-the-trainer model is being developed by Site 1. A cadre of instructors from various institutions will be trained to implement the program at their respective schools and to become master teachers who train and assist others in implementing the program. Instructors' guides and workshops are in the planning stages.

ATE Site 4 curriculum reform has led to development of multiple applications in the college's departments and divisions and has occasioned the involvement of instructors as trainers for their colleagues.

ATE Site 5's workshops have involved instructional personnel from 110 public schools and 23 community colleges. Of the 289 faculty members trained, 224 were from public schools and 65 from 2-year colleges. Fifty-four of the 82 counties in the

Table 1. Extent of Professional Development Implementation.

	Assessment Elements						
ATE Sites	Ongoing Training	Admin Support	Hands on Experience	Indiv. Training	Follow up Training	Mentoring	Train the Trainers
Site 1	X		X		X		X
Site 2	X	X		X	X		
Site 3	X		X	X	X		
Site 4	X	X	X		X		X
Site 5	X	X	X	X	X		X
Site 6	X	X	X	X	X		
Site 7	X	X	X		X	X	
Site 8	X	X			X		X
Site 9	X	X		X	X		
Site 10	X	X			X		
Site 11	X	X			X		
Site 12			X	X	X		
Site 13	X		X		X		

Table 2. Program Improvements as Reported by Sites 1–13.

Sites Outcomes	1	2	3	4	5	6	7	8	9	10	11	12	13
Course improvements	X	X				X	X	X	X	X	X	X	
Enhanced knowledge of technology	X	X	X	X	X	X	X	X	X	X	X	X	X
Increased industry understanding		X	X	X	X	X	X						X
Greater networking opportunities			X			X	X						
Student success	X	X		X	X	X			X	X	X		
Program sustainability	X	X		X	X	X							

state have been impacted by Site 5 teacher development efforts. Site 5 estimates that more than 60,000 public school students and twenty-five thousand 2-year college students have benefited from the professional development received by their instructors.

ATE Site 8 has expanded the effectiveness of their professional development training through a train-the-trainers process. An on-campus trainer (a faculty member) provides training to colleagues and brings in industry representatives to provide up-to-date instruction in technological fields.

A comparison of Table 1 and the Outcomes table (Table 2) on the following page demonstrates a possible relation between the seven assessment elements and self-reported program outcomes. Joyce and Showers (1983) have offered a clear and concise rationale for providing the professional development that incorporates the seven elements recommended in this paper: "If education programs are to be effective, they require sustained, continual training efforts that are adequately funded."

EFFECTS OF PROFESSIONAL DEVELOPMENT ON PROGRAM IMPROVEMENT IN ATE

While current research tells us that each of the seven elements of effective professional development is important – because each incrementally increases the likelihood of successful training – it is unlikely that developing training programs will be able to incorporate all elements in their programs' early stages. Indeed, the ATE projects bear witness to the fact that some good results can occur without all elements present (though, the more, the better).

In general, visited sites' participants describe positive results from the training they have received. The outcomes noted by project staffs and program evaluators reinforce the notion that professional development can have positive effects, even if all elements of effective professional development are not present at a particular point in time. In the more detailed site descriptions appended to this report it is clear that many sites are planning to add other elements to further improve their professional development efforts. Notable outcomes are presented in Table 2.

These data indicate that ATE *projects* have succeeded in developing professional training programs that provide faculty and students with improved learning, enhanced technology skills, increased access to technology fields, and expanded opportunities to sustain long-term professional development. Notably, they have done this without a template; therefore, there are varying approaches. Nonetheless, the seven elements of successful professional development can be used as a metric to systematically assess professional development as employed within ATE sites. Participants also reported that professional development was beneficial to their improvement efforts and provided a good example from which to build stronger professional development initiatives in the future.

FACING THE CHALLENGES AHEAD: CONCLUSIONS AND RECOMMENDATIONS

The ATE program was developed out of a need to bring educational institutions of the type represented in this study up to world-class standards in technology. The rapidly changing pace of technology throughout the industrialized world and the lack of institutionally sponsored professional development were primary reasons the ATE program was introduced. It is clear that NSF-ATE funds have been instrumental in producing higher levels of professional development, as described above. Without NSF-ATE funds, the extent and quality of professional development made available to faculties would likely not have occurred (site staff interviews).

Time remains a central issue for program participants. Most teaching staff members have to use their own time in the summer or during non-teaching periods to participate in professional development efforts (staff interviews). As indicated earlier, educational institutions value instructor-student contact hours most. Time for planning, professional development, assessment, and reflection of instructional effectiveness are considered to be the responsibility of instructors, but not the educational institutions for which they work. This thinking, which has evolved over the years, has often resulted in less than effective training programs. When budgets become tight, institutional support for professional development typically declines and less time is spent on upgrading the skills and capacities of the teaching workforce. The rationale is that it is the responsibility of faculty to keep up-to-date on both content and pedagogy in their areas of specialization. This line of thought frees the educational institution from direct responsibility for providing professional development to upgrade skills and competencies. It also contributes to the gap between the growth of technological development and the abilities of community colleges and other educational institutions to meet the technological workforce needs of their communities.

On the other hand, it is clear that when community college administrators begin to see significant increases in student enrollment, they become more interested in providing resources to the departments with growing classes (administrator interviews). Funds for updated equipment and faculty development are looked upon more favorably when there is more student tuition to pay for them. In one institution, when students complained to the college president that there weren't enough computers to support their classrooms, several new computers appeared the next day (student interviews).

Similar reports from administrators, faculty members, and students in many sites indicate that college administrations need to be less reactive and more proactive in keeping their college and faculty current in rapidly changing technological areas, as well as providing them with the instructional tools they need to be effective. ATE has demonstrated that when faculty are provided the opportunity to increase their skills and develop courses that meet the needs of the technological community, they rise to the occasion and become much more effective in carrying out their educational responsibilities.

Data gathered through site visits and an annual survey of projects by the WMU evaluation project (Gullickson et al., 2000) indicate that the ATE program has successfully engaged associate degree institutions and others in developing materials and programs and providing professional development services to help implement them.

Recommendations for introducing and/or improving professional development programs are easy to make, but often difficult to carry out. Nonetheless, the authors presume to offer three suggestions for institutions with ATE programs that are originating or evolving.

First, perhaps the most difficult is also the most important: The primary requirement for successful staff training is *institutional commitment*. Support from the organization's leaders is crucial to the development, implementation, and sustainability of effective professional development. If the leadership does not truly believe in the concept of continuing education for everyone, then surely any training program will fail for lack of resources and interest. Therefore, the administration must understand and exemplify the value of professional development. In essence, they must create the teaching/learning environment in which lifelong learning is a reality.

It is probable that more research on the value of professional development at the community college level would provide the persuasive evidence needed by institutional leaders to provide the necessary program support. Current research is largely focused on staff training on use of educational technology at the secondary level and, while it is helpful in considering the various aspects of professional development, it needs to be reinterpreted in light of the special needs of postsecondary schools and instructors.

Second, a continuing test of commitment for schools and colleges is adequate funding to support the time needed for instructors to learn, experiment, and implement. *Time and money*: the two principal barriers to professional development are the proverbial horse and carriage. We cannot separate them from one another if we want to make progress. In terms of time, McKenzie suggests that 5–10 days a year are needed for instructors to gain the information and achieve the confidence to use this information in their classrooms. That is a significant number of days and a sizable commitment of resources for educational institutions that currently may provide only a few days or afternoons each year to professional development.

Without such an investment, however, staff training is probably a waste of time and money. Expecting instructors to engage in voluntary and unrewarded training programs is unrealistic. Continuing education needs to be a required part of the work year that gives instructors the professional development they want and need (McKenzie, April 1991). It is essential that postsecondary institutions find funding for the one thing that will enable them to reduce the digital divide between a technology-based economy and the colleges'/universities' ability to train students for the 21st century workplace: That thing is professional development for teaching staffs. A possible solution to the time/money problems faced by the college is implementation of the train-the-trainers approach.

Third, the final recommendation is the need to *plan for assessment*. It is imperative to know where you want to go in order to get there. Seven basic elements that make up effective training programs have been used throughout this paper and may serve as a model for establishing or expanding professional development programs that result in enhanced teaching knowledge and skills. To repeat, those elements are: (1) ongoing learning and training; (2) institutional support; (3) hands-on and classroom-based experiences; (4) individualized training; (5) follow-up training; (6) mentoring; and (7) a train-the trainers approach to continuing education. It seems that we know what to do to achieve effective professional development; we must now define how to proceed at the community college level.

Despite the challenges we face in time, resources, and commitment, it is important that we persevere in providing quality professional development for instructors. To meet the demands of a rapidly changing society and economy, we need to think of continuing education as a substantial, long-term investment. That investment will bring us a committed teaching cadre and a well-prepared workforce. And that, after all, is what our efforts are all about.

NOTES

1. Much of the current literature on professional development, technology training, and technology integration addresses K-12 levels. While the elements of effective professional development span time and topic, it is apposite for the reader to consider that the community college perspective and faculty needs may differ.
2. The assessments in this section are based on self-reporting from the ATE site visits. Each site has noted the elements of effective professional development that it has incorporated into its training program.

REFERENCES

American Association of Community Colleges (2000). A look at the future. *AACC Research*. Washington, DC: AACC. Available online at: htp://www.aacc.nche.edu.

Arter, J. (2001, Spring). Learning teams – The way to go for professional development. *Classroom Assessment Connections*. Portland, OR: Assessment Training Institute, Inc.

Bents, R., & Howey, K. (1981). Staff development – change in the individual. In: B. Day (Ed.), *Staff Development – Organization Development* (pp. 18–33). Alexandria, VA: ASCD.

Brand, G. A. (1997). What research says: Training teachers for using technology. *Journal of Staff Development*, *19*(1). Oxford, OH: National Staff Development Council. Available online at: http://www.nsdc.org/library/jsd/brand191.html.

Browne, D. L., & Ritchie, D. C. (1991). Cognitive apprenticeship: A model of staff development for implementing technology in schools. *Contemporary Education*, *63*(1), 28–33.

Cook, C. J. (1997). *Critical issue: Finding time for professional development*. North Central Regional Educational Laboratory. Available online at: http://www.ncrel.org.

Darling-Hammond, L. (1994, November). *The current status of teaching and teacher development in the United States*. New York: Teachers College, Columbia University.

Darling-Hammond, L. (1991). The meaning of professional development in the 21st century. *Restructuring to Promote Learning in America's Schools*, videoconference #8. North Central Regional Educational Laboratory. Available online at: http://www.ncrel.org.

Dobbs, K. (2000, January). Simple moments of learning. *Training Magazine*, *37*(1), 52. Available online at: http://www.trainingsupersite.com.

Garmston, R. J. (1999). Better by the bunch. *Journal of Staff Development*, *20*(4), 64–65.

Guhlin, M. (1996). Stage a well-designed Saturday session and they will come! *Technology Connection*, *3*(3), 13–14.

Gullickson, A., Lawrenz, F., & Keiser, N. (2000). *Assessing the impact and effectiveness of the Advanced Technological Education (ATE) program: Status report 2*. Kalamazoo, MI: The Evaluation Center, Western Michigan University.

Guskey, T. R. (1999). Apply time with wisdom. *Journal of Staff Development*, *20*(2), 10–15.

Guskey, T. R. (1986). Staff development and the process of teacher change. *Educational Researcher*, *15*, 5–12.

Harvey, J., & Purnell, S. (1995, March). *Technology and teacher professional development*. Report prepared for the Office of Educational Technology, U.S. Department of Education. Santa Monica, CA: Rand Corporation.

Hord, S. M. (1997). Professional learning communities: What are they and why are they important. *Issues . . . About Change*, *6*(1). Austin, Texas: Southwest Educational Development Laboratory: Available online at: http://www.sedl.org/change/issues/issues61.html.

Johnson, M. A., & Johnson, G. A. (1999). The insiders: Development in school with colleagues can succeed. *Journal of Staff Development*, *20*(4), 27–29.

Joyce, B., & Showers, B. (1983). *Power in staff development through research in training*. Alexandria, VA: ASCD, 15–22.

Kimmel, H., Deek, F. P., Farrell, M. L., & O'Shea, M. (1999, May). Meeting the needs of diverse student populations: Comprehensive professional development. *School Science and Mathematics*, *99*(5), 241–243. Bowling Green, OH: School Science and Mathematics Association.

Mahoney, J. R., & Barnett, L., Editors (2000). *The learning edge*. Washington, DC: Community College Press.

McKenzie, J. (1991, April). Designing staff development for the information age. *The Educational Technology Journal*, *1*(4).

Occupational Outlook Quarterly (2000, Summer). *Futurework: Trends and challenges for work in the 21st century*. Washington, DC: Superintendent of Documents. Available online at: http://proquest.umi.com/pqdweb.

Persky, S. E. (1990). What contributes to teacher development in technology. *Educational Technology*, *30*(4), 34–38.

Shelton, M., & Jones, M. (1996). Staff development that works! A tale of four Ts. *NASSP Bulletin*, *80*(582), 99–105.

Sparks, D. (2000, March). Corporate lessons for evaluating staff development. *Results*. Oxford, OH: NSDC. Available online at http://www.nscd.org.

Sparks D. (1998, Winter). Using technology to improve teaching and staff development: An interview with Kathleen Fulton. *Journal of Staff Development*, *19*(1). Oxford, OH: NSDC.

Sparks, D. (1997, Fall). Reforming teaching and reforming staff development: An interview with Susan Loucks-Horsley. *Journal of Staff Development*, *18*(4). Oxford, OH: NSDC.

Sparks, D., & Hirsh, S. (1999). A national plan for improving professional development. NSDC Web site: www.nsdc.org/educatorindex.htm. (Look under "What's New.")

Spitzer, D. R. (1999, June). Embracing evaluation. *Training Magazine*, *36*(6), 42. Available online at: http://www.training.supersite.com.

Stager, G. S. (1995). Laptop schools lead the way in professional development. *Educational Leadership*, *53*(2), 78–81.

Sprinthall, N., & Sprinthall, L. (1980). Adult development and leadership training for mainstream education. In: D. Corrigan & K. Howey (Eds), *Concepts to Guide the Training of Teachers of Teachers*. Reston, VA: Council for Exceptional Children.

Tenbusch, J. P. (1998, March), Teaching the teachers: Technology staff development that works. *Electronic School*, 1–6. Alexandria, VA: National School Boards Association. Available online at: http://www.electonic-school.com/o398fl.html.

Watts, G. D., & Castle, S. (1993, December). The time dilemma in school restructuring. *Phi Delta Kappan*, *75*(3).

RECRUITMENT AND RETENTION

Arlen R. Gullickson and Gloria R. Tressler

INTRODUCTION

A primary goal of the National Science Foundation's (NSF) Advanced Technological Education (ATE) program is to increase the number of highly skilled technicians in the U.S.'s workforce. ATE uses community colleges as a "pipeline" for this purpose. As the term *pipeline* suggests, NSF and others view community colleges as a conduit for preparing technicians. This conduit provides educational experiences that prepare recruited individuals for highly skilled technician positions in businesses and industries. Recruitment – persuading individuals to enroll in a community college technician program – and retention – keeping students successfully enrolled in the program to completion – are essential to the success of the pipeline.

This paper addresses ATE projects' efforts to recruit and retain students in technological education programs. The paper's purposes are to:

- Briefly define and describe recruitment and retention as employed in colleges and especially ATE-based community college programs.
- Identify and describe factors key to successful recruitment and retention.
- Describe the expectations and actions of the ATE program, and those of funded *projects*,[1] in relation to the key factors for recruitment and retention.
- Identify and describe strengths upon which the program and its funded *projects* can build and further improve recruitment and retention.
- Identify weaknesses that may exist in the current program and make recommendations to effectively address them.

Evaluating the Upgrading of Technical Courses at Two-Year Colleges:
NSF'S Advanced Technological Education Program
Advances in Program Evaluation, Volume 9, 131–161
Published by Elsevier Ltd.
ISSN: 1474-7863/doi:10.1016/S1474-7863(04)09007-6

We divided the paper into three parts: (a) general context/background; (b) findings from our analysis of data;[2] and (c) a discussion of findings. The general context is divided into two parts. First, we describe findings from current literature that helped our understanding of the general background issues surrounding recruitment and retention and the various strategies employed to enhance both efforts. Much of this literature addressed higher education generally, but we paid special attention to those sources that focused on associate-degree institutions. From this literature, we identified three factors (goal commitment, preparation, and support) common to successful recruitment and retention practices. Second, we reviewed and briefly present the ATE program: Congressional expectations for the ATE program, the ATE program's expectations (guidelines) for funded *projects*, and key attributes of the program.

The findings we report come primarily from survey and site visit data gathered in 2000 and 2001. We construct and present a simple framework with three elements (Information, Preparation, and Support) to assess the ATE program efforts. Additionally, we present five case examples from survey and site visit findings to better show some of the methods used by *projects* to reach recruitment and retention objectives.

The discussion following the findings addresses implications for serving pipeline needs. We conclude the paper with several recommendations for further developing and improving the program.

THE RECRUITMENT AND RETENTION CONTEXT

General Background

Recruitment

The comprehensive nature of recruitment can be found in J. Smith's writing, "The primary goal of recruitment programs and activities is to influence the behavior of prospective students, their parents, and significant others in the college admission process ... [recruitment can include activities to]: (a) generate initial student inquiries and interest; (b) identify serious potential candidates; (c) convert prospects to applicants; (d) convert applicants to deposits; and (e) convert deposits to matriculants" (Smith, 1998).

The commonly known "bread and butter" methods of recruitment include college nights at high schools, scholarship offers, and brochures in the mail. Some other notable methods advocated in current literature include:

- Use of outside marketing consultants with 8% increase of enrollment noted (Lord, 2000).

- Collaborative inter-institutional practices for written materials, recruitment events, presentations, joint trips or visits, and electronic media with testimonials of success (Stonewater, 1999).
- Active recruitment of high-achieving Black students, especially through offers for paid tuition and room and board, book stipends, laptop computers, guaranteed summer internships or research jobs, and all-expense paid weekends. There is evidence of enhanced reputations for Historically Black Colleges and Universities with more National Achievement Scholars, at times, than Princeton or Harvard (Roach, 2000).
- Use of nontraditional marketing techniques to appeal to Generation Xers with evidence of drastic improvement of name recognition for the institution (16.6–94.5%) and first increase in enrollment in seven years (Raisman, 1999).

Some writers used logic models to explain the process students follow in choosing a college. These models provide both a progression of steps and specific factors known to affect recruitment success. Hossler and Gallagher (1987) described a widely accepted three-stage student enrollment behavior model. This model divides the decision process into: (a) college aspiration; (b) search and application; and (c) selection and attendance. Coupled to these three stages, Belcher and Frisbee (1999) list 17 factors that influence decisions at these three stages. The factors (as listed by the authors and specific to an automotive program) include: (1) friend(s) at university/community college or high school; (2) reading this university's catalog; (3) high school/community college counselor/teacher; (4) parent(s)/relatives; (5) alumni of this university; (6) reputation of automotive program; (7) technology recruitment activities; (8) university recruiters visiting my high school; (9) athletic advisor/coach; (10) admission office at this university; (11) campus visit; (12) reputation of the university; (13) university recruiters visiting my community college; (14) community in which university is located; (15) bulletin board advertising at my previous school; (16) promotional materials (brochures, letters, videos); and (17) articulation or direct transfer from community college.

Most high school graduates are recruited to higher education. Approximately three-fourths of high school graduates enroll in colleges and other forms of higher education either immediately upon high school graduation or within two years of graduation (Haycock & Huang, 2001).

Retention

Retention can be defined as "... the maintenance of students' satisfactory progress toward their educational objectives until the objectives are attained" (Dolence, 1998, p. 89). The objective may be a college degree, completion of a program (course sequence) or even a single course required to obtain a job or a pay raise. Especially for community college students the educational objective is likely to be "shorter term" than degree attainment.

Retention is closely coupled to recruitment. Much like the chicken or egg proposition, there can't be retention without effective recruitment. Yet, effective retention attracts applicants. Additionally, "... retention rates [are] critical measures of institutional effectiveness" (Moller-Wong et al., 1999). Indeed, graduation rate, the sine qua non of retention, is often used as a performance measure for institutions and programs.

Efforts to improve retention rates typically center on freshmen orientation, academic advising, and continued financial assistance. But, information on best retention practices is sparse. Authors tend to focus more on the difficulty of obtaining good retention information than on methods to improve retention.

Three factors contribute to misunderstandings of retention. First, researchers often presume that all students seek a degree. In those situations students, who complete their personal objectives short of a degree and discontinue enrollment, are treated as "drop outs." This yields a depressed retention rate value. Second, terminology is not consistent across writers – a wide array of terms (e.g. persisters, graduates, attainers, transfers, stop-outs, dropouts, and dismissals) is used to refer to students and their enrollment status. Third, students may be completing programs or degrees at various institutions, with discontinuous enrollment, and over an extended period of time. These enrollment variations necessitate longitudinal studies, such as the National Longitudinal Survey (NLS), to accurately determine completion rate (Tinto, 1993).

Despite the difficulties several aspects of retention seem clear.

- Attrition is greatest at the freshman level and declines in following years. Levitz and Noel (as cited in Moller-Wong et al., 1999) put the attrition rate in the first year (students leaving their current institution) at about a third overall. Haycock and Huang (2001) noted substantial differences between 2- and 4-year colleges. They found that more than a quarter of those in 4-year colleges and nearly half of those in 2-year colleges do not make it to the sophomore year. The rate of attrition declines by about 50% every year thereafter.
- Fewer than half of those enrolled graduate. Several studies put graduation rates in the range of 40–50%. Moller-Wong et al. (1999) found that approximately 45–50% complete their undergraduate degrees. Similarly, for 4-year institutions, Erickson and Strommer (1991, p. 41) found that over the past 20 years, the rate of graduation nationally has been roughly 40% in 4 years, not quite 50% in 5 years. For 2-year institutions, Tinto (1993, pp. 18–19) found completion rates for full-time students across a 3-year time frame to be 43%.

- Overall attrition rates have remained fairly stable for the past 60 years (Moller-Wong et al., 1999).
- Attrition is largely due to dropouts rather than forced dismissals.

Key Factors for Recruitment and Retention

Most recruiters argue that they do not strive for a 100% retention rate because not all students may be the right fit for the institution (Rummel et al., 1999). However, we argue that stronger preparation and information programs should provide better initial "fits" for recruits, which should result in higher retention rates. Concomitantly, the reputation and backing of retention programs that offer good quality supports should yield higher recruitment numbers.

The studies and essays we reviewed identify three general factors as important to individuals' decisions to initially enroll and their subsequent decisions to stay enrolled to completion of their planned objective.

- Personal commitment to the goal.
- Academic preparedness.
- A sufficient support base.

Several studies note the importance of educational aspirations and goal commitment. Tinto (1993) and Cope and Hannah (1975, as cited in Moller-Wong et al., 1999), especially note the importance of commitment to the academic or occupational goal. Tinto states "The commitment of the student to the goal of college completion had the strongest positive effect on the decision to remain in school." He continues, "Given sufficiently high goal commitment, individuals may decide to 'stick it out' even in unsatisfactory circumstances" (1993, p. 43). Hurd (2000) noted the importance for students to know what they are working toward so that they don't flounder, become frustrated, and leave.

Studies also confirm the importance of addressing those in a position to affect the aspirations of the prospective or enrolled student (i.e. parents, teachers, counselors, academic advisors, employers, administrators, and peers) (Belcher & Frisbee, 1999; Munro, 1981). Munro noted that parents as a group had the strongest direct effect on students' goal commitment (1981). In today's society quite likely housemates or spouses will have a similarly strong direct effect.

Academic preparedness provides educational and work opportunities. The importance of academic preparedness is well known and understood, but not necessarily appreciated by those who most directly influence students (i.e. family, friends, and peers). For young persons whose parents have not completed college or do not hold skilled technology positions, awareness and interest may not be generated in the home. For these individuals, it is especially important to create awareness and interest early in the educational process (e.g. at the middle school level) and for them to gain requisite skills for entry to college. Mulder (1991) concluded that lack of preparedness presents the most severe problems for minority students. These preparedness problems appear to stem from factors such as inadequate curricula, linguistic difficulties, and family concerns. These findings suggest the necessity of special efforts with minority students to assure that they gain knowledge and skills requisite for admission to technology-based college programs.

When the objective is larger than a course or a short sequence of courses, students require a sound support base to stay in school and maintain progress toward program completion. The support-base element captures several variables of importance. Studies commonly highlight a student's financial status as a major consideration for entry to college and staying in college to completion. This appears to be especially important to retention beyond the freshman year with student dropout often related to family financial problems (Hurd, 2000). Hurd reports that students often stop-out when they meet financial difficulties. The student will leave and work for a year, then come back (p. 44). Two recent studies (Roach, 2000; "UNCF," 2000) especially emphasize the importance of financial assistance from collaborating business and industry partners for technology-based programs. These programs often directly serve minority groups.

Johnson (1997) studied retention rates for commuter students. He found that retention rates for these students are consistent with findings from studies conducted of noncommuter, residential students at other institution. His findings suggest that retention issues are consistent across types of students. His study has particular value because the characteristics of commuter students are much like those of typical associate degree students. That is commuter students usually have responsibilities and pressures such as full or part-time jobs, and family and home responsibilities. These responsibilities and pressures differ from traditional campus-based students.

Several studies note institutional climate as an important support factor for students. It has been noted that most students drop out of school for nonacademic reasons such as personal, social, and financial (Cambiano et al., 2000; Kalsner, 1991). With regard to minority students, Parker (1997) noted the importance of creating "a campus atmosphere where students are presented with a mandate to succeed, not the right to fail" (p. 1). This is also especially true for students with disabilities. Adapted physical facilities and assistive technology devices are examples of an institution's willingness to cultivate an institutional climate that supports *all* students (Malakpa, 1997). Tinto summarizes the matter of support well. He states, "we must . . . avoid the tendency to assume that all members of a particular group have the same interests or needs. Though it is sometimes necessary for institutions to develop programs targeted to the needs of distinct groups of students, it is always the case that program action must be guided by the assessment of *individual* needs" (1993, p. 181).

Some writers, like those cited above, provide individual gain statistics to provide an indication of the impact of a strategy or technique tried. We found no experimental studies (i.e. studies employing a control group) on this topic.

THE ATE PROGRAM

When the U.S. Congress passed the *Scientific and Advanced Technology Act of 1992* as Public Law 102–476, two incorporated purposes were "To establish a national advanced technician training program, utilizing the resources of the Nation's two-year associate-degree-granting colleges and to *expand the pool* of skilled technicians in strategic advanced-technology fields . . ." [italics added].

> The ATE program flows from and responds to the Congressional mandates. Work of the ATE program, in turn, is based upon its published guidelines. To understand ATE expectations for recruitment and retention, NSF-ATE Program proposal guidelines were reviewed for fiscal years 1994–2002. The guidelines were searched for their specific references to recruitment and retention and to underrepresented and/or nontraditional students.[3]

This review of guidelines revealed the following:

(1) The guidelines encourage efforts that will increase recruitment and retention, especially by ATE centers.
(2) The guidelines have been evolving. This is most notable in the expanded wording with regard to underrepresented and/or nontraditional groups of students.
(3) Recent years have seen a shift to more directness about the issue of accountability. For example:
  - Although proposers were always asked to lay out plans for recruiting, now regional centers for manufacturing or IT education (FY 2001) are being asked for mechanisms for measuring numbers of students recruited, numbers retained until competencies are achieved or certifications received, and the numbers of those who partake in internships, etc.
  - Fiscal year 2002 guidelines add "Reporting Requirements" of responding to a survey requesting information about the number and characteristics of students affected by a *project's* activities.
(4) Proposers are also being asked (FY 2002) to work with industry partners to address "capacity building (recruitment, retention, and placement of students)" by collaboratively engaging in various activities with students at secondary and college levels.

(5) Due to Executive Orders 12876, 12900, and 13021, guidelines under the heading of Budgetary Information for fiscal years 2001 and 2002 have changed by removing the requirement for cost sharing/matching by Historically Black Colleges and Universities (HBCUs), Hispanic Serving Institutions (HSIs), and Tribal Colleges and Universities (TCUs).
(6) Although not referenced in the guidelines as recruitment-and retention-related work, several specific methods were suggested that should ultimately lead to more success in these areas (e.g. recommendation to use modern instructional technologies, articulation of courses and programs, professional development).

Locus of Activities

Congress named associate-degree-granting institutions as the primary site for technician training. Additionally, it is repeatedly stated in NSF literature that these institutions should play a leadership role in ATE *projects*. For this paper, we focused on these "community colleges." However, it should be noted that several recruitment and retention methods cited in this paper may have overlapping application to the three primary points of activity for the ATE program: (a) middle and high schools; (b) community colleges and to a lesser extent baccalaureate colleges; and (c) collaborating business and industry locations.

Key Participants

Three key groups participate in the recruitment and retention of ATE students:

- Business and Industry – Business and industry have substantial needs for well-educated and skilled technicians. This key group provides career opportunities and is responsible for establishing entry level and cutting-edge skill and knowledge requirements for careers. This group is increasingly involved in sponsoring all levels of recruitment and retention activities in collaboration with schools.
- Schools – This group, primarily associate-degree-granting colleges and other educational institutions that work collaboratively with them, provides direct educational instruction and programming to develop skilled technicians to serve workforce needs. In addition, this group provides support and information for students.
- Students – *Prospective* students are the individuals being recruited to the identified technology careers. These recruits include: (a) those at the secondary school

level who are making initial decisions about career options and opportunities; (b) employees in business and industry seeking to continue their education to increase their options and opportunities; and (c) to a lesser extent, students at the college level who have not yet made career decisions or who are considering changing their career paths. *Enrolled* students are those individuals who have matriculated and are actively engaged in a technology-based program.

ANALYSIS AND FINDINGS FROM SURVEY AND SITE VISIT DATA

We constructed a framework of three elements – Information, Preparation, and Support – to organize the data and synthesize findings. This framework was based on our review of literature and ATE materials and matched to the identified key factors (personal goal commitment, preparation, and support). Both the literature and the ATE work suggest that success requires attention to all three elements. That is, a well-balanced recruitment and retention program enrolls and retains students who are adequately:

- Informed about career opportunities and a college's ability to provide a sound educational route to those opportunities.
- Prepared to succeed in the program.
- Supported by an educational/institutional system in order to reach a successful completion of the program.

The following more detailed "operational" descriptions are provided to clarify our basis for categorizing information provided in survey responses and site visit reports.

Information

This element is based on the expectation that sound information regarding educational and vocational options helps to develop and sustain goal commitment aspirations. Although not able to alter the student-related factors directly, college recruitment and retention activities can modulate the effect of these factors by equipping the students, or those who influence them, with accurate and helpful information. Information objectives work in tandem with preparation and support objectives by addressing both preparation requirements and support resources available to students.

For the ATE program, the key aspect of this element is creating an interest in an advanced technology career and the college educational program as a means to that career. The element of Information should communicate the local, regional, or national need for graduates in a specified field with concomitant career and benefit opportunities.

Provided through all stages from pre-enrollment through program completion, the element of Information identifies work-based opportunities and educational/vocational goals. It focuses on developing and sustaining goal commitment on the part of the student. It also identifies procedural and educational steps needed to attain those goals. Much like basic marketing principles, this element provides potential customers (prospective students) and those in a position to influence them (e.g. teachers, parents, and peers) with increasing levels of awareness about and interest in the product. It is also concerned with retaining repeat customers (enrolled students) by maintaining their level of interest and motivation about the product (chosen advanced technology field) at high levels.

Preparation

The element of Preparation is needed to create a solid educational and experiential foundation. Students use this foundation to make informed decisions and reach their potential. It is not sufficient for students to be informed only of career opportunities and routes to those opportunities. They must also prepare adequately to navigate the path to their chosen careers. This element:

- Is an educationally based element that must begin early for success in any field, not just advanced technology careers.
- Requires solid grounding in science, math, and technology along with a bedrock of literacy.
- Includes preparation through various experiential methods (internships, work experience, or work study).
- Applies also to those who can influence students – for example, teachers and guidance counselors must continue to keep their skills sharp and be properly prepared to educate and guide prospective and enrolled students.

Support

This element focuses on creating and maintaining an environment to attract and keep students. Setting the stage with necessary props – if you will – for success. The primary objective here is solving needs-based issues so that interested students know they can obtain sufficient support to enroll in the program and continue to completion. Examples of needs-based issues include the financial viability of attending college; family responsibilities and family care options; social integration within the college setting; student advising; and the need for Americans with Disabilities Act (ADA) accommodations.

Financial needs are met in a variety of ways, for example, by providing direct support to visit a college campus, scholarships, student loans, paid work experiences and internships, and other aids such as free or reduced costs for child care while attending courses. Institutional (college based) and industry scholarships serve as a primary tool to address financial need. Similarly, as noted in the ATE program description, inter-institutional collaborative and articulation agreements provide a variety of mechanisms to support students.

Though more difficult to measure, support also includes other factors:

- Providing work and study environments that are attractive to and supportive of prospective students (when recruiting) and of enrolled students (when retaining). The environments should be consistent with the work-based opportunities the students seek to obtain.
- Answering supply issues such as availability, scheduling, and locating classes to meet student needs.
- Maintaining attractive, modern educational facilities with state-of-the-art laboratory and instructional equipment.
- Sustaining a solid reputation through high standards for teaching, content, and certification or graduation from the program.

ATE Survey and Site Visit Findings

Using the Information, Preparation, and Support elements as categories, we:

- Reviewed the Survey 2001[4] and site visit data to identify and describe the nature and extent of activities conducted to serve recruitment and retention objectives.
- Separately analyzed responses to the Survey 2001 recruitment (Program Improvement section question 13a) and retention questions (Program Improvement section question 13b) to assess patterns in the use of Information, Preparation, and Support efforts.
- Analyzed site visit data[5] both to gain a general sense of whether these data and survey data provided consistent results and to identify exemplars of recruitment and retention practice.

These data provide evidence of substantial recruitment and retention efforts by *projects*, individually and collectively. Tables 1–3 identify and list the various

Table 1. Objectives Served and Methods Employed to Address the Element of Information.

Objective: Make prospective ATE students aware of, knowledgeable about, and interested in pursuing careers in advanced technology fields.

Sample methods

- Career fairs
- Presentations by college representatives
- Presentations by industry representatives
- Tours of college programs or industry
- Career opportunities matrix
- Job shadowing
- Field trips/site visits to industry
- Brochures, catalogs
- CD-ROMS, Web-based audios, videos and printed materials, presentations (e.g., providing overview and interview with industry reps)
- Web sites
- Billboards/radio/newspaper/magazine articles or advertisements
- Conferences on minorities in technology
- Orientation courses
- Contacts through partnerships with local, regional, or government organizations (e.g., public housing authority, native organizations)
- Mass mailings to pools of students–postal or electronic
- ATE project-sponsored national high school competitions

Objective: Keep prospective or enrolled ATE students informed of requirements and steps that must be taken to achieve entry to the career, including basic requirements for entry to and completion of a college degree and/or requirements for successful completion of courses and industry or program certification.

Sample methods:

- Career advisement via school counselors or teachers
- Preadmission counseling – especially for older adults
- Extended video or other instructional options
- Industry-college articulation programs or agreements
- College-secondary school articulation agreements
- Presentations to workers at job sites

Objective: Convey institution's ability and willingness to address needs-based considerations (e.g., financial, academic skills, family responsibilities) so that interested students know they can obtain sufficient support to enroll in the program.

Sample methods:

- Identify needs (e.g., financial, academic skills, family responsibilities) through needs assessment activities such as surveys; focus groups; interviews with and input from counselors, teachers; family planning; and other community service agencies.
- Provide information about these matters (scholarships, student loans, education-employment opportunities, etc.) and ways support needs can be met through brochures, career days, counselor assistance training, and other dissemination media.

Table 1. *(Continued)*

- Describe access and support for students with disabilities and provide information describing/verifying nature and extent of access and support available.

Objective: Make those who are influential to prospective or enrolled ATE students aware of and knowledgeable about educational opportunities and careers in advanced technology fields.

Sample methods:

- Technical awareness workshops for professionals
- ATE *project* faculty personal visits with high school teachers/counselors
- College counselor training
- ATE laboratory open houses for teachers and administrators
- Technology fairs for community members, including parents and peers
- Industry tours for teachers at all levels of education
- Presentations for parents about course and dual credit options
- CD-ROMS, Web-based audios, videos and printed materials, presentations (e.g., providing overview and interview with industry reps)
- Billboards/radio/newspaper/magazine articles or advertisements

methods employed and represent responses from the majority of *projects* (not all *projects* engaged in recruitment and retention activities). We found that most methods group into several subcategories corresponding to *project* objectives or intended outcomes. We used those subcategories to organize data within the respective tables.

Participant responses tabulated by element(s) addressed for recruitment and retention are attached to this paper.[6] The summary results from those tables are provided in Table 4.

As Table 4 shows, 45 *projects* gave us usable/categorizable responses to the Survey 2001 Program Improvement section question 13a, which asked respondents to briefly describe specific steps taken to *recruit* students to their program. The majority of recruitment efforts appear to focus on providing information to prospective recruits.

Because the literature points out the importance of external influencers (e.g. parents and teachers), we separately categorized Information and Preparation activities in terms of who was the recipient of the information. Nearly all respondents' activities (93%) could be categorized as work to provide information to students *directly*. Some of the *projects* (26%) *indirectly* provided information by informing others (e.g. teachers or counselors) who work with the students. Only a small minority reported methods strongly tied to either Preparation or Support (approximately 10% per category). The one *project* reporting indirect preparation provides educational experiences for teachers.

Data for retention findings in Table 4 come from a single question that asked respondents to describe specific steps taken to *retain* students. Most respondents

Table 2. Objectives Served and Methods Employed to Address the Element of Preparation.

Objective: Create a solid *educational* basis upon which prospective and/or enrolled ATE students can make informed decisions and reach their potential.

Sample methods:
- Workshops/seminars
- Tech prep courses/introductory career classes
- STEM instruction in middle and high schools
- "Bridge" or developmental courses to address academic underpreparedness
- Study skills instruction
- Articulated secondary school/college courses

Objective: Create a solid *experiential* basis upon which prospective and/or enrolled ATE students can make informed decisions and reach their potential.

Sample methods:
- Summer tech camps
- ATE-related work-study
- Industry-based student internships
- Hands-on interactive jobs for a day
- Real-world, industry-based problem assignments
- Technical Olympics

Objective: Keep teachers and counselors prepared to educate and guide prospective or enrolled ATE students.

Sample methods:
- Industry-based faculty internships
- Faculty/teacher ATE workshops with certification or other educational credit available
- Technical skills upgrading professional development workshops
- ATE *project*-led activities for special education teachers
- Hands-on interactive jobs for a day

listed a few (1–3) methods, and only one project reported many (9) different methods. As Table 4 shows, the majority of reported methods (82%) address the element of Support with relatively few – 22 and 18%, respectively – providing information and preparation-type activities.

While the site visit reports provide substantial detail, they did not clearly establish whether the *project* methods served one objective, recruitment or retention, or were intended to serve both. As a result, the findings for recruitment and retention are combined in the Table 4 summary of site visit results. Additionally, we chose to include all possible recruitment and retention activities identified in site visit reports (i.e. activities were included in our tabulations even if not designated under recruitment or retention headings). The results show that 10 out of 13 *projects* engaged in activities addressing all 3 elements of information,

Table 3. Objectives Served and Methods Employed to Address the Element of Preparation.

Objective: Offer solutions for a student's personal considerations.

Sample methods:

- Accommodations (including attitudinal) for physical or learning needs
- Personal/psychological counseling
- Peer tutoring/college-provided free tutoring
- Positive and encouraging staff and administration
- Mentoring programs
- Programs for those returning to workforce (i.e., homemakers)
- Small class sizes with opportunities for better interpersonal relationship with teacher and classmates

Objective: Offer solutions for a student's practical considerations.

Sample methods:

- Scholarships, grants
- Convenient campuses, transportation, and course times
- Articulation agreements allowing dual credit at secondary and postsecondary levels and/or enabling easy transfer of course credit among postsecondary institutions
- Job placement
- Child-care facilities
- Laboratories open for extended hours/Saturdays
- Online courses

Objective: Establish and maintain high quality teaching and content standards to support effective recruitment and retention by ATE programs.

Sample methods:

- Up-to-date industry-aligned course materials
- Cutting-edge laboratories and equipment
- Sufficient supply of equipment for all students
- Programs well regarded by the community
- Faculty available for individual student assistance
- Technical educators to support and advise teachers in rural areas

preparation, and support. That finding appears to be consistent with survey data for the categories of Information and Support, but is larger than the survey results for the category of Preparation.

Case Examples

Tables 1–4 identify a wide array of recruitment and retention activities, but they do not provide many details that show the full power of the various methods. To more clearly depict the particular emphases of *projects* and their attention to Information, Preparation, and Support aspects, five case examples are presented.[7] These cases show recruitment and retention from quite different vantage points.

Table 4. The Number of ATE Projects' Recruitment and Retention Efforts as Reported in Survey 2001 and Site Visit Reports.

Source	Element(s) Addressed				
	Information		Preparation		Support
	General or Direct[a]	Indirect[b]	General or Direct	Indirect	
Survey 2001 program improvement section					
Number of *projects* listing recruitment efforts for item 13a ($n = 45$ *projects*)	42	12	5	1	5
Number of *projects* listing retention efforts for item 13b ($n = 49$ *projects*)	9	2	8	1	40
Site visit results					
Reported items related to recruiting/retaining students ($n = 13$ *projects*)	10	7	9	12	11

Notes: Any one *project* may have reported several specific activities or methods that fit into an individual category. A *project* was counted only once, regardless of the number of activities it listed.

[a] General or Direct indicates activities that have either a general application to multiple audiences or to prospective ATE students only.

[b] Indirect indicates activities that are intended to reach those who are in a circle of influence to the prospective ATE student.

One shows the strong relationship of an ATE *project* with the needs, influence, and support of business and industry and its effect on student recruitment. A second shows a comprehensive, systematic approach by a community college that focuses most heavily on retaining students in the program. The third describes community college students serving as mentors to high school students, effectively melding technician programs in secondary schools with an associate degree program. The fourth addresses use of community organizations to recruit and retain underrepresented groups. The fifth provides an abbreviated analysis of a *project* that has not met recruitment expectations and factors that have deterred its recruitment success. All directly quoted information is from the WMU evaluation project's site visit reports or survey responses, with permission from principal investigators related to the cited *projects*. In keeping with commitments made in obtaining access to these sites, site-specific names are omitted and referred to by number (e.g. Site 1).

Site 1. Welch and Gullickson (2001) described how this college engages in a variety of collaborative efforts with industry. Four examples are briefly described. The first shows industry involvement in awareness and preparation activities. The remaining three show recruitment via special types of collaboration, each with a specific goal and identified group of individuals. In each instance, the industry partner is integral to the educational program offered and participates fully from the inception of the idea through completion of the process. This report is paraphrased below.

(1) Awareness and preparation activities
In collaboration with Nortel, a summer tech camp was held that focused on wireless technology. Two 3-week camps were held. Fifty-seven students completed them. Approximately two-thirds of these students received their amateur radio licenses during or after the camp. Nortel provided $34,000 and nearly 60 mentors to help support the camps. Other corporate sponsors included Southwestern Bell, Kimley-Horn, and Sunbelt Plastics, Inc. Separate workshops and seminars have been developed and presented to middle school teachers, students, and parents both at two additional locations, a college and a community college

(2) Industry certification courses
This community college has become one of the major certification sites in the U.S. for Cisco Systems, the world's largest producer of Internet hardware and software. Cisco developed the curriculum, and this site has been authorized to offer the classes (Cisco calls them academies). The site is one of six institutions in the nation to offer advanced Cisco Systems training leading to a Network Professional certification. Currently, some 700 students are enrolled in Cisco classes.

(3) Technician retraining
The College offers off-campus programs in conjunction with several industries including Alcatel, MCI World Com, ST MicroElectronics, and Southwestern Bell. These businesses and industries support participation of their employees to update technician skills. Through a series of courses participants, gain requisite technician skills in hardware-based telecommunications industry. The program prepares participants to move to higher skilled positions in the industry. The courses are provided at the industry site or campus depending on availability of equipment and course instruction facilities. The College is currently offering 20 on-site classes that enroll about 300 students. A full-time staff person has been hired by the Division of Engineering Technology to respond to these industry needs.

One collaborative enterprise is a program of studies for employees of an individual company. This program provides a sequence of 10-week courses, followed by 2 weeks off for a period of 2.5 years. Most instruction occurs on Saturdays. The instructor travels to the collaborator's site to provide direct instruction. Students come to the College to use its equipment for required course labs. The collaborator pays tuition and book costs for participants. Most participants in this program started as production floor workers. Some now work in manufacturing hardware (equipment to provide telephone dial tone). The program is directly targeted at development of technician skills and is quite challenging.

(4) Engineer retraining
The college was approached by industry to train degree-holding engineers. The company provides the facilities and software needed and pays employees while they participate in the class. This is not a degree program but rather an effort to "re-tool" the engineers. The program is expected to continue for three to five years.

Site 2. This community college developed an extensive 4-phase system that Sterry and Schwabenbauer (2001) outline in their site visit report for the ATE evaluation project. This report has been edited to accommodate updated information provided by the *project* PI. Edits are enclosed in brackets [].

Phase One – During the first semester that students are in the program, they are introduced to a series of new project-based, competency-driven student-learning modules that introduce career options and the work of engineering technical professionals. These learning projects are set in a "hands-on" environment that simulates authentic industrial learning situations. Students are introduced to "softer skills" such as working in teams, written and oral communication, and learning to analyze and solve real business problems.

First semester students also learn how to use basic industrial equipment in the laboratory settings at the new Reynolds Technology Center. This new $8.1 million facility, which opened in mid 1998, houses many of the laboratories and classrooms that are used to deliver Engineering Technologies Division programs. These modern facilities are clean, orderly, and extremely well maintained. This environment provides a positive atmosphere for student learning and is representative of laboratories in modern technology focused industries.

The Phase One experience is, in part, intended to motivate students to select an area of study and stay in school. The college just completed its first semester of this new approach to learning for Technology students, and based on discussions with students, the initial feedback on Phase One and the entire program was very positive. [Student retention has improved well above the current rate across the College.] Since many students tended to leave school after their first semester, the expectation is that this class will help to motivate and encourage students to continue on to graduation. Another factor that may have contributed to the higher retention rate is a more careful screening of incoming students.

Phase Two – After the initial core semester when all the students take a similar program and have learned team building, communication skills, critical thinking, and problem solving by working in an industrial laboratory environment, they select a major field of specialization, but spend much of the next three semesters working on core content that ranges across specializations. Students study in areas such as computer-aided design, CNC, electronics, fluid power, material fabrication, manufacturing, quality, and control systems.

Again, in these Phase Two classes, students work individually and in teams. The real advantage is that during their first semester core classes they learned how to work with team-based activities. As a result, they are capable of learning technical specialties and functioning in an industrial work environment.

Phase Three – This is a single semester employer-paid internship. The College carefully screens employer programs to be sure they are providing high quality learning experiences for the students with opportunities to formally integrate academic competencies with a work experience.

There is very careful monitoring of student performance during the internship experience. Students must apply for a position and be recommended by a faculty member. They submit a resume to the employer and go through the same interview process that the company uses for all applicants. A school intern advisor works directly with the student intern and the employer in planning, monitoring, and assessing the results of the internship experience. Advisors also visit students on site to monitor performance throughout the internship.

Employers also agree to rate the student performance after the internship in areas such as quality of work, technical skills, people skills, communication, individual initiative, and contribution to company goals.

Phase Four – In this Capstone experience, usually in the last semester of the program, students select and work in teams on a project that often comes from their internship experience. They work to solve real industrial problems. This experience provides for application and reinforcement of knowledge, skills, and attitudes learned earlier in the program. The overarching outcome of the Capstone experience is to engage students in applied research, identify best practices in the laboratory, and deploy the most promising back to business and industry.

During the Capstone project, students learn to select team members with different skills required to deal with the problem. For example, computer-aided design skills are needed for conceptualizing and designing system solutions, electrical and electronics skills for the controls, and manufacturing skills for mechanical and quality assurance. As in any industrial or business environment, students learn that it is not just our individual performances and knowledge that matter, but that it is the combined ability of a team with the right mix of skills that really makes the difference in solving real-life problems.

Industrial partners have focused on outreach by helping to market the benefits of the program to the communities in which they operate, including middle schools and high schools of the region. A special industrial advisory committee has been established to work with marketing the program. Industrial partners have also been helpful in providing scholarships and internship opportunities for students.

Outreach and targeted recruitment have been directed at high schools and middle schools to advertise the new program. Perhaps the most innovative effort was the summer science, math, and technology academy initiative for students from area schools. [In the academies, 8th–10th grade students spent 2 weeks on campus during the summer and had opportunities to participate in teamwork and multiple career field experiences.] The Project believes this is a particularly important time to reach students to make them aware of career opportunities that are available to Engineering Technologies Division graduates.

Students are provided the usual counseling, tutorial, and advisement assistance. With the new program of integrated communication, mathematics, and science, students find learning to be meaningful because of the application to technology projects. The broadly based, hands-on experiences provided in Phase One of the program help students to be successful as evidenced by the improved retention rates.

Administrators and faculty members believe that better student selection will help to improve student success by creating a better "fit" between the program and student abilities, interests, career goals, and learning styles.

Site 3. This project offers a unique approach to keeping students informed, prepared, and supported. The site visit report by Lavoie et al. (2000) gives a detailed presentation of this system of recruitment and retention. This report has been edited to accommodate updated information provided by the *project* PI and specific college names have been removed. Edits are enclosed in brackets [].

[The] . . . primary goal of the project is to create an articulated environmental technology curriculum at the secondary level that adds to the already established environmental technology four-year degree, two-year degree, and year-long certificate programs offered at the community college [and two collaborating collegiate institutions]. [This] . . . ATE project intends to meet this goal through three related goals. The first goal is curriculum development and the creation of articulation agreements between the Community College and the high schools served by the College. The second goal is to develop and implement a mentor/mentee program that provides high school students in the articulated environmental technology curriculum with field experience as they shadow community college students in internship experiences. The third goal supports the curriculum development, articulation agreements, and mentoring program through a summer professional development session called "Summer Institute 2000" and follow-up workshops.

We [site visitors] believe that the project also has the goal of strengthening relationships with local high schools and industry and, ultimately, of recruiting more students to the Community College.

According to documents and the PI, faculty, and administrator interviews, other collaborating institutions include businesses, industries, and town and state government offices in the area. Individuals from these institutions serve on an advisory board related to, but not exclusively for, the project. The board includes representatives from environmental technology industries in the area, high schools, and four-year colleges. The board primarily guides [the community college's] course content, suggests courses to offer, and also makes recommendations for secondary-level tech prep course content. Many of the institutions on the board are also internship sites for [the community college's] environmental technology program and some, therefore, have hosted high school mentees. The PI also regularly collaborates with local institutions, which may or may not be members of the board, as she coordinates the environmental technology program internships. She uses these to guide tech prep course curriculum decisions and to market the mentor/mentee program. There are over [50] internship locations (eight of which were involved in the mentor/mentee program). The project also collaborates with an ATE center, by using the competencies developed by the center for environmental technology education to guide the tech prep course development.

The following description of the mentoring program is based on interviews with the PI, past and present mentors and mentees, and the Internship and Shadowing Coordinator. Mentors, who are internship students at [the community college] for its environmental technology program, are trained by the [the community college's] mentoring expert from the Business and Industry Council. Mentors then describe their internships to the high school juniors in the environmental technology program, and the school matches mentor/mentee pairs based on common work interests. Each mentor typically has two mentees. The mentors guide their mentees in completing a class research project in environmental technology, and they spend at least one day together at the mentors' internship sites. The relationship lasted one semester the first year and has been changed based on mentor and mentee feedback to last from October to May this year. At the end of the mentoring experience, all mentors and mentees participate in and exposition where students present their projects. Mentors are paid a stipend from the grant.

All three-project goals help the community college recruit new students into its environmental technology programs by spreading the word about the programs to high school students and offering them college credit for articulated high school courses.

The ATE grant provided the funding for the project's professional development activities – the four-day Summer Institute and its follow-up meetings. The project goal of providing a professional development experience to explain the mentor/mentee program and articulation agreement process for the three tech prep courses to the local high school teachers was met.

As a result of the ATE funding, the project established the mentor/mentee program, placing fifteen high school students with eight mentors the first year and placing ten students with five mentors the second year. The high school students received advising on their environmental technology research projects and visited the mentors' internship sites as a result of the program.

The strengths of the mentoring program are its advising impacts on the mentees and leadership impact on the mentors. The mentees consistently state that the experience helps them with their projects but mostly helps them learn more about their interests and the jobs available in different areas of environmental technology and other sciences. A few new mentees expressed the hope that they will learn more specific techniques in the field as a result of working with their mentors at their internship sites. The students were mature and well spoken as they discussed their participation and learning in the program. The mentors reported that the program gives them the opportunity to share their knowledge and expertise with others and to expand the understanding of environmental technology education in the community.

The mentoring program is successful because it is integrated into the curriculum, with a research project required for a class as result of the mentoring experience. The teachers support and promote the program and have invited mentors to visit with their mentees at the school during class-project time this year. The mentors are committed to making a difference in the lives of students and to sharing their excitement about their environmental technology internships. Although they are paid a stipend, two of the mentors donated their stipends to scholarship funds last year, and one mentor is trying to procure a laptop computer for his mentee who is a disadvantaged student.

Internship sites mostly support the exposure of mentees to the industry as well. The mentoring program also benefits from the already established internship program at [the community college].

Site 4. This project demonstrates a dedication to traditionally underrepresented groups of students for advanced technology fields. The ATE survey respondent described this *project* as serving students who are "virtually all . . . disadvantaged in some way" (2000). His words from the survey are captured here.

> The program relies on community organizations for students for the program. This is essential because many of the participants have not been well served by formal institutions of education and would not be responsive to recruitment by community colleges. The community organizations also provide case management and support services, which are critical to helping students stay in what is to them a very rigorous program. The community organizations work with the community college to help program graduates find jobs with local companies and enroll in associate degree programs at the college. They also provide follow-up support to ensure that students are advancing both in their education and careers.
>
> The staff of the partner community organizations provides intensive case management to each participant in the program. This intensive one-on-one support is critical to enabling these educationally and economically disadvantaged students to stay in the program.
>
> The community college and community organization together have built very strong relationships with local employers. Representatives from local companies were extensively involved in the development of the curriculum to ensure that it meets their needs and continue to monitor the program's performance and make suggestions for improvement. The community college organizes special seminars and other events for students in the program to introduce them to the college's technical programs and assist students with placement testing, GED preparation and financial aid applications before the students complete this developmental program.

Site 5. This project serves an isolated region of the country, and had an especially compelling story/problem. This project is funded for three years and is now in its second year. As proposed, the project will develop a specialized technician associate-degree program. In developing his request for NSF funding, the project director obtained commitments of interest from across the state and strong institutional commitments as well. Yet at the beginning of year 2 of the grant, only 4 persons have enrolled in the 20-opening program. The enthusiasm and hard work of the project director, the large number of commitments by others, combined with the small number of enrollees, raised questions about recruitment. Why is it that such efforts go "unrewarded?" Why is there not a waiting list of persons hoping to enter the program? The visit by Gullickson to the site addressed those questions and brings to light some but certainly not all the underlying causes. At the of the site visit, no project evaluation work had been initiated. Contractual problems had delayed initiation of the evaluation by several months. Additionally, follow up of the project's recruitment efforts is not a part of the evaluation expectations.

Here are some of the factors elicited from the project director:

(1) Commitments made by the institution were not fully honored. For example, the director was promised instructional space for the program, but when the project was approved the space had already been assigned to a different program. This made it necessary for the project director to spend considerable time and energy obtaining and preparing other space options.
(2) This program built upon several existing courses but required development of a large number of new courses. By the time of the site visit, the project director

had prepared 10 courses and received institutional approval for their use. He is currently preparing an additional 4 courses; bringing the total new direct- and laboratory-instruction courses to 14. Each course requires substantial development time and effort. Each must go through a prescribed institution approval process, which includes such things as preparation of a course syllabus, identification of viable textbooks, departmental approval, etc. These courses must be approved as part of approval of the full program. Of course, a course must be approved before it can be taught. Currently approved courses (i.e. courses in existence before start of the project) also need to be modified or replaced by newly developed courses to fully fit the new program's degree requirements.

(3) Many students who desire to enter the program do not have the requisite skills. Here the difficulties are more than "skin deep." That is, for many or most of the interested students their skills are not sufficiently strong for direct entry into the program. Additionally, for many, their skills are sufficiently lacking that they cannot be corrected through a single course or brief summer program (Two of the four who entered the program in year two are not well prepared and are struggling to succeed). Rather, these students need much stronger math skills, which must be developed through high school programs. In the project director's perception, the students' math deficiencies are sufficiently large to require pointing to this type of program at the beginning of high school so that students obtain the requisite background through a regular course of study in high school. This requires a substantial change in students' understanding and concomitantly counselors' and teachers' orientations at the middle school and high school levels.

(4) The program prepares students for a new occupation, one that is not currently listed in the state's directory for employment. This has direct implications for recruitment in that the job listing is not available in materials regularly shared by the state offices with high schools, employers and others. That the occupation is not listed stems from the fact that this specialized industry's demands are changing. In part, the work is now being done either by unqualified persons or those who have 4-year baccalaureate degrees. Approval of a new category seems vital to changing industry habits and bringing the option to employers and employees. It is also vital as a recruitment tool to show employment possibilities for graduating students and encouraging high school students to obtain proper mathematics preparation.

(5) The project is minimally staffed. Almost all responsibility falls on the project director. He has handled planning, development, and most implementation with some instructional help by other professionals for short courses. He also obtains assistance from up to four associate-degree student assistants. Adding

to the weight is the fact that the project director must also learn new methods and techniques in order to teach the newly approved courses. The program has received approval for hiring a new staff member. However, the approval is for one year only, with continuation contingent upon viability of the program. The new staff member cannot be brought on board before the third and final year of the project. As such, it serves as a springboard for continuation, but the person will not arrive in time to provide much needed assistance in the developmental efforts for the project.

(6) Recruitment to the program did not begin in a timely manner. Courses were developed and approved for the program's first year before recruitment efforts began. Information was therefore sent late to high school counselors and others who could effectively direct students' attention to the program and assist/support their preparation for and application to the program. Information was not delivered until near the end of year 1 and/or the beginning of year 2 of the project. Then, other events (e.g. September 11, 2001) distracted people from the message. Additionally, at the time of Gullickson's meeting with the project director, steps had not been taken to follow up with the recipients of the message to determine if the information had been received and effectively used, and whether there were other steps that might be taken to improve the viability of the message for increasing student applications to the program.

DISCUSSION

As stated in our opening paragraph, the path from school to employment in business and industry is commonly referred to as a pipeline. Traditionally, a generic educational pipeline process could be viewed in the following way: Secondary schools play a role at the beginning of the pipeline, preparing prospective students; associate-degree-granting and baccalaureate institutions recruit and retain these students in their educational and training programs; and business and industry step in at the end of the pipeline as employers of graduates.

Applying this concept to the ATE program, we consider the pipeline as a conduit for engaging students in an educational and training process that prepares them for positions in technologically intensive work in ATE-related positions in business or industry.

However, even though the process remains the same in some ways, it has changed significantly in scope and complexity. Secondary schools are no longer the sole source of prospective students. Now, for technology-based programs, there are multiple sources from which individuals are recruited into this pipeline

– secondary school students, current business and industry employees, college students undecided on their careers, and adults seeking reentry to the workforce. Even though associate-degree-granting institutions continue to provide the core educational programs for ATE fields and as such continue to play the primary role in recruitment and retention issues, collaborative arrangements (e.g. articulation agreements) may involve various parties. For instance, students may earn credit for college programs while still in secondary schools or while employed in business and industry. This increases their exposure and experience with ATE material and facilitates access to an advanced technology degree.

Persons already employed by business or industry may have the opportunity to move into more skilled positions as they progress through the educational process of technology education programs. In that regard, placement too has changed from being simply the hiring of persons to fill technology positions to one in which many current employees are "relocated" to positions requiring greater technology skills. Business and industry are no longer entrenched solely at the output end of the pipeline. Instead, they actively engage in recruitment and retention efforts for their employees as well as other prospective employees. As this description suggests, the boundaries of the pipeline have blurred as secondary schools and higher education overlap at the recruitment-retention boundary, and business and industry now overlap at all stages. We believe the methods employed by ATE projects for recruitment and retention reflect these pipeline changes. The details of our five case examples demonstrate the degree of involvement of business, industry, and other community organizations in the process of technology education.

The 3-stage enrollment model of Hossler and Gallagher (1987) expanded with 17 factors by Belcher and Frisbee (1999) seems viable when considering the decision process as simply one of college choice that is made by students enrolled in secondary schools. However, today's prospective ATE students are a heterogeneous group. This group requires an enrollment model that: (a) addresses student entry into a program from a work-based situation; (b) includes participation by business and industry representatives as key stakeholders in the process; and (c) focuses clear intentions to inform, prepare, and support students seeking advanced technology degrees or certifications.

Our survey and site visit data bring forward important information for projects that seek to develop and/or improve their recruitment and retention efforts. Tables 1–3, for example, list many methods that projects can employ to serve these objectives. The case examples additionally provide extended descriptions of several methods.

The survey and site visits have quite different bases and therefore provide different information. The survey sample is census like and provides data from nearly all funded projects in place for a year or more. The site visits were conducted at a judiciously selected set of 13 projects to represent both good examples of project work and diverse characteristics of the projects. The survey findings, therefore, more accurately describe the full population and indicate the nature and extent of work that projects identify as recruitment and retention. The site visit reports, though less viable as descriptors of the population of projects, add support and substance to the survey through their contextual information and extended descriptions of project activities.

For two of the elements, Information and Support, survey and site-visit findings were compatible. That was not true for Preparation. Our site-visit findings show more recruitment and retention efforts for Preparation than were reported by the projects' survey respondents. Since our site visit to 13 projects identified nearly as many Preparation methods as reported by all projects in the survey, it appears that our analysis gives greater credit to this category than do projects themselves. This is probably due to our purposeful inclusion of professional development as an indirect preparation element that supports recruitment and retention. Additionally, we believe the reputation of well-trained and knowledgeable educators is a strong attractant for students.

Several aspects of survey and site visit findings suggest the viability of the ATE program's recruitment and retention efforts. First, projects engage in recruitment and retention activities that are consistent with those recommended in the literature. As such project efforts seem appropriate and likely to serve well (i.e. they have face validity). Second, in similar fashion some projects report increases in their student populations. By itself it is not proof, but it does suggest the program is having an impact. Third, the projects have developed substantial numbers and types of collaborations with business and industry partners. Business and industry support from these relationships likely improves students' understanding of technicians' roles in these fields and increases job opportunities for persons who complete the respective certification or degree programs. These points are not sufficient to claim success for the programs' recruitment and retention efforts. Both comparative studies and longer term trend data will do much more to buttress the current tentative indicators.

Most of the reported recruitment efforts focus on the element of Information, with limited work applied to Preparation or Support elements. In contrast, most retention efforts address Support, with limited attention to either Information or Preparation elements. When looked at as a single recruitment/retention entity, these findings suggest that project actions parallel our own expectations that recruitment will focus more heavily on providing information, and retention will give greatest attention to support, with preparation (which is also the goal of the program) maintaining a visible presence throughout.

Finally, site visit information indicates that many recruitment and retention activities coincide with NSF-based efforts but are not part of the funded NSF projects. As such it can be argued that the recruitment and retention achievements are not due to NSF support and NSF should not lay claim to them (e.g. in reports for the Government Performance and Results Act [GPRA]). Yet, it is clear that projects can and do leverage the coincident efforts to serve project-based purposes. In these regards both the local community college and NSF objectives appear to benefit.

CONCLUSIONS

(1) NSF's guidelines for funded *projects* consistently encourage the development of programs that recruit and retain groups that have been traditionally underrepresented in technology positions. Stronger and more specific language regarding accountability has been added to the guidelines in recent years.

(2) ATE program efforts address all three parts of the recruitment and retention model we presented. Its multifaceted goal may be articulated in the following manner:

- Interest prospective students in ATE courses, certifications, or degree programs.
- Keep ATE teachers, staff, and advisors trained and prepared to educate and support students.
- Prepare prospective ATE students through educational and experiential opportunities.
- Offer avenues of support that meet the student's needs to such an extent that the enrolled student persists in the courses, and certification or degree program until ultimate employment or advancement in an ATE field.

(3) Many activities are being employed to recruit and retain students in ATE-funded instructional programs. These methods appear to be substantial and likely to produce good results. Especially noteworthy are the many collaborative arrangements between colleges and business and industry. These arrangements are key for the success and survival of the ATE program. In addition to being integral to a *project's* materials development and sustainability, in some cases the arrangements are directly responsible for the recruitment and retention of substantial numbers of students. This may include outreach to nontraditional students by bringing educational programs to employees and thereby upgrading their technology knowledge and skills.

(4) Concomitant to use of the recruitment and retention methods is the importance of strategic planning to employ these techniques in timely and effective ways. As the one case example shows, actions delayed may result in substantially lowered applications to the program.

(5) Testimonials of success and promised success can be found throughout the survey and site visit data. We found many exemplars (e.g. 350 middle and high school students visited campus on one day, or the course enrollment has doubled) that testify to the productivity of various activities. Generally, however, *projects* have not reported systematic evaluation of their success in recruitment and retention.

(6) It was not always clear whether recruitment/retention activities were supported by or otherwise linked to the ATE grants, or if any linkages existed between ATE *project* efforts and other community college and business/industry efforts.

(7) Within the ATE program we identified a broad range of activities that we see as affecting recruitment and retention. This range of activities is important because recruitment and retention are not only interrelated, they are inseparable from the primary functions of all educational institutions. Not surprisingly, site visits show these interrelationships and functions much better than do survey findings.

(8) Both the literature and *project* data suggest that the effectiveness of each R & R method may be enhanced by an array of factors. These factors include the timing and timeliness of the approach, the interpersonal nature and extent of the contacts, and the level of expertise to which the student is exposed. In other words, how a message or service is delivered, when, how often, and by whom, all will play significant roles.

RECOMMENDATIONS

(1) Further strengthen, promote, and clarify recruitment and retention objectives in the ATE guidelines:
- Strengthen ATE guidelines to encourage all *projects* involved in direct ATE student education to build strong recruitment and retention programs.
- Promote the development of recruitment and retention as a system of interrelated activities that keeps all stakeholders (i.e. students, employers, educators, administrators) aware of responsibilities, opportunities, and outcomes.
- Clarify recruitment and retention as a system rather than individual activities. Emphasize that it is integral to the ATE program to encourage matriculation in and successful completion of the college's technology program. State that the R & R system should provide traditional and nontraditional prospective and enrolled students with needed elements of Information, Preparation, and Support.

(2) Require evaluation to document recruitment and retention efforts and accomplishments and/or to identify ways for improvement.

- Establish key questions and indicators for use by internal or external *project* evaluators to:
  - (a) Increase accountability for funds expended (by confirming that recruitment and retention efforts are productive).
  - (b) Gain information that can serve recruitment and retention efforts. For example, brief surveys can be tailored to identify students' support needs.
- Consider including evaluative methods such as the following:
  - (a) Use longitudinal tracking approaches to better understand the retention rates of ATE students. For example, class size for courses can be tracked across terms or years. A random sample of students can be identified on entrance to the program and their progress tracked as they work toward completion of personal objectives. Such tracking efforts need not be extensive or onerous to provide valuable feedback to the project.
  - (b) Use student focus groups to keep ATE projects in line with student expectations and vice versa.
  - (c) Identify student objectives as part of student application data to plan for success and to encourage students' persistence.
  - (d) Use a specific time frame and breakout retention and completion rates by major types of student objectives (e.g. degree, program, or course sequence).
  - (e) Use a checklist of commonly understood recruitment and retention activities on future surveys to better assess project status.

NOTES

1. NSF funds projects and centers. The convention *projects*, in italics, will be used to denote both projects and centers as a group, unless specifically indicated otherwise.

2. Data sources: Three primary types of data sources were employed in composing this chapter:

- NSF-based publications, especially ATE guidelines from fiscal years 1994–2002.
- Published literature on the topics of recruitment and retention.
- Data from two years of this ATE evaluation project's work including.

- Results from two annual surveys of current ATE *projects*, conducted in spring 2000 and winter 2001.
- Information reported from site visits to 13 *project* sites conducted in the late 2000 and early 2001.

3. These references are available from the author, please ask for Appendix A. Additionally, in that appendix we noted if a citation was directed to ATE projects, centers, or generally stated.

4. Because 96% of the sample in Survey 2001 also participated in Survey 2000, which asked the same questions, we chose not to analyze data from the 2000 survey. The analyses used items 13a and 13b of the Program Improvement section of the survey. We did, however, include some Survey 2000 responses as illustrative for this paper.

5. We were not able to clearly distinguish and categorize activities as recruitment or retention based.

6. All appendices are available from the lead author. Appendices B and C are constructed from the Survey 2001 data, while Appendix D is prepared from the site visit data. Appendix B addresses recruitment efforts; Appendix C addresses retention; and Appendix D, because the site visit reports combined recruitment and retention descriptions, addresses recruitment and retention together. These appendices in table format show which elements receive attention and the extent to which individual *projects* comprehensively address all three identified elements.

7. Three of the examples are drawn from site visit reports; one is taken from Survey 2000 data, and one is prepared from a separate site visited because of apparent recruitment problems.

REFERENCES

Belcher, G. G., & Frisbee, R. L. (1999, December 12–15). Factors that influence students to attend four-year automotive programs. Paper Presented at the annual meeting of the Association for Career and Technical Education, Orlando, FL.

Cambiano, R. L., George, S., & DeVore, J. B. (2000, Winter). College student retention at a Midwestern university: A six-year study. *Journal of College Admission*, *166*, 22–29.

Dolence, M. G. (1998). Strategic enrollment management. In: C. C. Swann (Sr. Ed.) & S. E. Henderson (Ed.), *Handbook for the College Admissions Profession* (pp. 71–91). Westport, CT: Greenwood Educators' Reference Collection (Published in association with the American Association of Collegiate Registrars and Admissions Officers).

Erickson, B. L., & Strommer, D. W. (1991). *Teaching college freshmen*. San Francisco: Jossey-Bass.

Haycock, K., & Huang, S. (2001, Winter). Are today's high school graduates ready? *Thinking K-16*, *5*(1), 3–17.

Hossler, D., & Gallagher, K. S. (1987). Studying student college choice: A three-phase model and the implications for policymakers. *College and University*, *62*(3), 207–221.

Hurd, H. (2000, October 26). Staying power: Colleges work to improve retention rates. *Black Issues in Higher Education*, *17*(18), 42–46.

Johnson, J. L. (1997, September). Commuter college students: What factors determine who will persist and who will drop out? *College Student Journal*, *31*, 323–332.

Kalsner, L. (1991, Fall). Issues in college student retention. *Higher Education Extension Service Review*, *3*(1), 1–7.

Lavoie, B., Igoe, A., & Keiser, N. (2000, December). *ATE evaluation project final site visit report: ATE Project in Environmental Technology Education*. Kalamazoo: Western Michigan University, The Evaluation Center.

Lord, E. (2000, May 19). Community colleges turn to consultants to help them recruit and retain students. *Chronicle of Higher Education*, *46*(37), A65–A66.

Malakpa, S. W. G. (1997, Summer). Problems in the admission and retention of students with disabilities in higher education. *Journal of College Admission*, *156*, 12–19.

Moller-Wong, C., Shelley, M. C., II, & Ebbers, L. H. (1999, Fall/Winter). Policy goals for educational administration and undergraduate retention: Toward a cohort model for policy planning. *Policy Studies Review*, *16*(3/4), 243–277.

Mulder, A. E. (1991, Summer). Minority student enrollment. In: D. Angel & A. Barrera (Eds), Rekindling minority enrollment. *New Directions for Community Colleges*, *74*, 31–37.

Munro, B. H. (1981, Summer). Dropouts from higher education: Path analysis of a national sample. *American Educational Research Journal*, *18*(2), 133–141.

Parker, C. E. (1997, February 20). Making retention work. *Black Issues in Higher Education*, *13*, 120.

Raisman, N. A. (1999, February/March). Leave the field of dreams! Successful strategies for marketing the community college. *Community College Journal*, *69*(4), 14–19.

Roach, R. (2000, October 26). Battling for the best. Black schools experience a renaissance in recruiting high-achieving students. *Black Issues in Higher Education*, *17*(18), 36–41.

Rummel, A., Acton, D., & Costello, S. (1999, June). Is all retention good? An empirical study. *College Student Journal*, *33*(2), 241–246.

Smith, J. E. (1998). Recruitment: Student outreach strategies. In: C. C. Swann (Sr. Ed.) & S. E. Henderson (Ed.), *Handbook for the College Admissions Profession* (pp. 127–139). Westport, CT: Greenwood Educators' Reference Collection (Published in association with the American Association of Collegiate Registrars and Admissions Officers).

Sterry, L., & Schwabenbauer, A. (2001, January). *ATE evaluation project final site visit report: High performance technicians in distinctive manufacturing: An innovative approach*. Kalamazoo: Western Michigan University, The Evaluation Center.

Stonewater, B. B. (1999, Summer). Collaborative admissions and recruitment practices. *New Directions for Higher Education*, *27*(2), 45–50.

Survey Respondent (2000, July). Responses to ATE evaluation survey program improvement section questions 11a, 11b, and 11c.

Tinto, V. (1993). *Leaving college: Rethinking the causes and cures of student attrition* (2nd ed.). Chicago: University of Chicago Press.

UNCF gets $1 million from Lockheed Martin for tech campaign (2000, December 7). *Black Issues in Higher Education*, *17*(21), 32.

Welch, W. & Gullickson, A. (2001, March). *ATE evaluation project final site visit report: Advancing careers in technology and science*. Kalamazoo: Western Michigan University, The Evaluation Center.

SUSTAINABILITY AND LONG-TERM IMPACT

Frances Lawrenz and Nanette Keiser

THE CHALLENGE OF DEFINING SUSTAINABILITY FOR THE ATE PROGRAM

According to the dictionary, sustainability is the ability to prolong or to supply with sustenance. This straightforward definition takes on a much more complex character when considered in relation to the Advanced Technological Education (ATE) program because of the diverse nature of this program. The first complexity is that the ATE program operates through several mechanisms called "drivers." These include collaboration, materials development, recruitment and retention, program improvement, and professional development. Projects can be funded to work with one or all of these drivers. Therefore, what would "be prolonged or supplied with sustenance" and how it would be done varies dramatically project by project. As another element of complexity, awards are made in two categories – projects and centers. Although all *projects*' (i.e. projects and centers) ultimate goal is to support the development of well prepared technicians, projects tend to focus on only one or two of the ATE program drivers, while centers typically address most or all of the drivers. Additionally, centers tend to receive larger grants to match their expanded scopes of work, always receive funding for multiple years, serve as model programs for other institutions and organizations, and disseminate information to a region (e.g. several states or the nation as a whole).

Evaluating the Upgrading of Technical Courses at Two-Year Colleges: NSF'S Advanced Technological Education Program
Advances in Program Evaluation, Volume 9, 163–184
Published by Elsevier Ltd.
ISSN: 1474-7863/doi:10.1016/S1474-7863(04)09008-8

Setting these complexities aside, in a simple sense, sustainability for the ATE program could mean continuation of whatever activities (drivers) had been supported by the NSF grant, including institutionalization. This is consistent with the definition given for sustainability by the Community College Research Center (CCRC) in their study of the ATE program as well. They defined sustainability as "The state where the major activities involved in the ATE program continue even after the grant expires." This continuation could take on several forms, including those listed below:

- Retaining all grant activities for a *project* but lowering work scopes for these activities.
- Moving forward with a subset of a *project's* activities.
- Supporting the former grant activities by the organization(s) that received the NSF grant.
- Receiving funding from other sources external to the institution via commercializing materials or processes (e.g. professional development), obtaining other grants, or obtaining support from collaborators.
- Disseminating a *project's* activities or products through their continuation or use at different institution(s).

The CCRC provides a similar list of examples and groups them into two approaches: to become self-sufficient and to integrate ATE into the college organization. Naturally, outcomes or processes that are not successful or of high quality should not be sustained. This places a burden on the ATE *projects* and NSF to determine where efforts for sustainability should be focused.

This leads us to consider what NSF's perception of sustainability is and why NSF considers sustainability important. NSF is less entrenched than other federal agencies, but it also has somewhat lower and less consistent levels of funding. It has capitalized on its flexibility to move quickly in funding innovative, cutting-edge programs. It is also interested in obtaining the most leverage it can with the money it has. In the Education Directorate in particular, NSF has a history of funding novel/creative approaches or model demonstration programs. It has been careful to not lead its grantees into expecting long-term NSF support. Most grants have been short term (i.e. 1–3 years), and only recently have longer time frames been considered. Even within these longer periods, however, the expectation is that once whatever was proposed is accomplished, demonstrated, or implemented, NSF will no longer be involved, and the continuation of quality outcomes is the role of other agencies or the states.

Because of this history of "in and out" funding, the notion of NSF supporting sustainability of projects or their effects is an evolving issue at NSF. It is becoming increasingly important because of the Government Performance and Results Act's

(GPRA) focus on outcomes, even though the Fiscal Year 2001 Final Revised Performance Plan does not formally address sustainability. NSF wants to be able to show that its funding produces long-term, continuing effects. This is particularly true in the Education Directorate where sustained improvement is desired for the nation's approaches to science and mathematics education.

Although NSF staff members talk about sustainability, it is not a theme commonly written about by the Foundation. For example, a search for "sustainability" using the general search function on the NSF home page (http://www.nsf.gov) produced 193 hits. However, many of the hits were duplicates, and only 10 were related to sustainability in education. One of these was in the summary of the meeting on the Science and Technology Center (STC) directors' workshop. This report, *Building Bridges*, states among other issues, "Plan for sustainability of the STC from the beginning; Build a transition team that focuses mainly on sustainability issues." The October 2000 Grant Proposal Guide and the document *A Guide for Proposal Writing* do not mention sustainability. Furthermore, sustainability was only mentioned in 9 of the current 73 education program announcements located at http://www.nsf.gov/home/programs/ehr.htm.

The ATE program is housed in the Division of Undergraduate Education (DUE) of the NSF Education Directorate. The DUE Annual Report for Fiscal Year 2000 contains only two references to sustainability – one under the Evaluation of the Institution Wide Reform Initiative with the Course and Curriculum Development (CCD) Program (p. 44) and the other under the publication description for DUE 0085870: Columbia Pubscape – A Core Integration System for a National Science Digital Library Publishing Center (p. 60). In 1998 and 1999, sustainability is mentioned in the program announcement for DUE under the Collaboratives for Excellence in Teacher Preparation (CETP) project.

The program solicitations for ATE mention sustainability only briefly and only in the more recent solicitations. Sustainability is not mentioned in any of the ATE program announcements through the year 2000. In the 2001 ATE Solicitation, "sustainability" appears only once, under "V. PROPOSAL PREPARATION AND SUBMISSION INSTRUCTIONS, A. Proposal Preparation Instructions, Full Proposal, Project Description." It states, "The Project Description should explain the project's motivating rationale, goals, objectives, deliverables, and activities; the timetable; the management plan; the roles and responsibilities of the PI, co-PI(s), and other senior personnel; *the prospects for sustainability after the period of NSF funding*; the evaluation plan; the dissemination plan; and results from prior NSF support" [italics added].

The information provided above shows that NSF's interest in sustainability in relation to the ATE program is evolving. There is not single definition of sustainability, and those that do appear are subject to change. In other words, what is expected from ATE *projects* in terms of sustainability is evolving. Additionally, ATE *projects* funded earlier are learning about the need for sustainability retroactively.

A LITERATURE-BASED SUSTAINABILITY CHECKLIST

Because there is no official definition of sustainability, we turned to the published literature. In order to review it, we searched a variety of databases and used several terms. We obtained several hundred references, but only a limited number were relevant to the ATE program,[1] which were the references on the sustainability of organizational change. Copies of these books or articles were obtained and read. The following is a brief summary of articles providing advice or models on how to sustain change. Generally, these articles were advocacy statements based on the authors' experiences with sustaining change rather than empirical studies.

Howard and Howard (2000) presented the Self-Determination/Reliance Model (SDR). This is based on psychosocial development and associated with the monitoring and assessing of the processes and outcomes of community groups. Critical movement on the identified dimensions indicates progress toward sustainability. These dimensions are accountability; decision making; information; knowledge and skills; and resource mobilization. Flower (1996) described five fundamentals for organizational change: husbanded resources, abundant relationships, abundant information, distributed power, and a clear sense of purpose. Schwartz (1994) listed several guidelines for managing and sustaining change. He stated that successful change requires employee participation, training, provision of continuous feedback, a reward system, and development of group norms. Gabelnick et al. (1990) offered a checklist for sustainable learning communities in colleges. The list includes determining focus and design; using faculty resources; coordinating with current institutional initiatives; obtaining and maintaining administrative support; having needed resources; using promotion and marketing; having institutionalization of the concept; and using program improvement mechanisms.

In addition to these various lists, articles referenced the need for organizational cultural compatibility and the ability of programs to create and command value. Schneider et al. (1996) reported that the cultural aspect suggests that change is not sustained when: (1) the change is inconsistent with existing climate and culture; and (2) changers fail to build a climate and culture to support the change. They go on to state that changes are more likely to be sustained when the organization is ready and prepared for the change and when the change permeates all levels and functions of the organization. Cropper (1996) suggested that the creation of value falls into two domains: (1) consequential value such as legitimacy, security, and efficiency; and (2) constitutive value, which is based on efforts being valued (i.e. expressions of purpose and fit with institutional context, capacity, and conduct). For example, in the ATE program, consequential value could be provided in terms of legitimization from the business/industry collaborations or the use of standards. Security could be through the use of existing programs and professors and emphasis on how the new programs will increase the security of their jobs and the security of the institution in terms of new students.

Taken together, these offer a fairly comprehensive set of issues to be addressed or elements necessary for successful sustainability. These issues were consolidated into the following seven elements and form the basis for our comprehensive, literature-based sustainability checklist.

(1) Wide Participation and Clear, Shared Purpose (i.e. abundant relationships, clear sense of purpose, employee participation, clear focus and design, development of group norms).
(2) Abundant Information Available and Used to Improve Program and Reward Effort (i.e. abundant information, accountability, use of program improvement mechanisms, continuous feedback, reward system).
(3) Abundant and Needed Resources, Resource Mobilization.
(4) Knowledge and Skills/Training.
(5) Decision Making/Distributed Power.
(6) Coordination with Current Initiatives, Administrative Support.
(7) Use of Promotion and Marketing/Husbanded Resources.

A SUCCESSFULLY SUSTAINED ATE PROJECT

Although this checklist drawn from the literature is helpful in defining sustainability in more concrete terms, it may underemphasize the richness of the entities it is intended to represent. Therefore, we constructed a description of an imaginary, successfully sustained project that participates in all the drivers to illustrate what might be reasonable expectations for such a project. To help highlight how this description relates to the checklist, we have referenced specific elements within this scenario.

The XYZ technician education program is healthy. The new faculty member who was hired during the ATE grant has received a permanent appointment, and we have a firm commitment from industry partners for a continuing budget for upgrading our equipment on a rotating basis (3. Abundant Resources/Resource

Mobilization; 6. Administrative Support). This is smaller than the budget we had under the ATE program, but we still can plan ahead.

We have solid enrollment. The recruitment procedures we put in place under the ATE grant are not as elaborate, but the contacts facilitated by the grant are still in place. We have an articulation agreement with several school districts. Through that agreement we interact with the local K-12 school districts, especially the counselors, and the science, mathematics, and technology teachers to help guide students to our program (1. Wide Participation; Shared, Clear Purpose). In some cases, students can take courses for simultaneous credit for high school graduation and college credit, and in other cases, the high school students can receive advanced standing at the college. We also have several mechanisms in place to support the students once they come to our institution. We have 5 years of external funding from a local business to support a summer mentoring placement where students from our college get paid to work at this business (3. Abundant Resources/Resource Mobilization). Additionally, the people providing the generalized support services are well informed about our program and its needs so that they can tailor the resources to our students (1. Wide Participation; Shared, Clear Purpose; 3. Abundant Resources/Resource Mobilization; 5. Decision Making/Distributed Power; 6. Coordination with Current Initiatives/Administrative Support; 7. Use of Promotion and Marketing/Husbanded Resources).

We are engaged in the development of materials to help improve our instruction. One of the pieces of curriculum we designed turned out very well, and we commercialized it (3. Abundant Resources/Resource Mobilization). Profits from the sale of these materials are used to keep it current and to support more local development work. We want our materials to incorporate more in-depth understandings of the science and mathematics principles behind the more technical concepts. We also want to help our students become more adept with workplace skills such as communication and group work. With the very small amount of money now available, materials development is more incremental. It generally involves trying to improve or update at least one section of a course every time it is taught. We all try to help each other by critiquing materials, and we routinely gather student outcome data against which we can ascertain the effects of any changes. The feedback from our advisory committee also highlights needed changes or effective elements of our instruction. Work on improving course instructional materials is considered in the merit pay schedule (1. Wide Participation; Clear, Shared Purpose; 2. Abundant Information Available and Used to Improve the Program and Reward Effort; 5. Decision Making/Distributed Power).

We have an advisory committee of representatives from the several employers of our students. This advisory committee meets only once a year now, but we communicate electronically at other times, especially when an important question comes up (1. Wide Participation; Clear, Shared Purpose). An important part of our meetings is when the business/industry people inform us about the on-the-job levels of performance of our graduates using evidence they have gathered throughout the year (2. Abundant Information Available and Used to Improve the Program). We then discuss what all of us can do to make the graduates even better prepared (1. Wide Participation and Clear, Shared Purpose; 3. Resource Mobilization; 5. Decision Making/Distributed Power).

As part of this discussion with business/industry, we arrange for short exchanges of faculty and business/industry people to promote understanding which are paid for by industry (3. Abundant Resources/Resource Mobilization; 4. Knowledge and Skills/Training). We discuss arrangements to attend the national meetings of the related businesses/industries (7. Use of Promotion and Marketing/Husbanded Resources). We also consider future directions of the businesses/industries and make modifications in our program in anticipation of these changes (3. Resource Mobilization; 6. Coordination with Current Initiatives). We also have some members from our local K-12 school districts and the nearby 4-year college on our committee. These educational institution members also function as a subcommittee in which we work on guaranteeing the smooth entry of high school graduates into our program and the transfer or continuing education of those who wish to pursue education at a 4-year institution (1. Wide Participation; Clear, Shared Purpose; 6. Coordination with Current Initiatives). We have continued the development of the articulation agreements begun under the ATE grant, and several decisions have been made. These contacts also aid in the recruitment and retention of students.

Although we meet less often than we did when the ATE grant was in operation, the faculty teaching the basic science and mathematics courses interact with us on a regular basis to ensure that what they are teaching meets our needs (1. Wide Participation; Clear, Shared Purpose; 6. Coordination with Current Initiatives). They share their student assessment results with us so we have input into what sort of knowledge should be assessed and what levels of accomplishment we require (2. Abundant Information Available and Used to Improve the Program; 3. Resource Mobilization).

We try to engage in professional development. There is very little money available for it without the ATE grant so we compensate by piggybacking on other opportunities (4. Knowledge and Skills/Training; 7. Husbanded Resources). As mentioned above, faculty members spend time in local businesses/industries while their counterparts do some instruction. When we hire adjunct instructors with cutting-edge expertise, we all try to attend their classes. As part of the hire, the adjuncts present a seminar to the rest of the faculty (1. Wide Participation; Clear, Shared Purpose; 3. Resource Mobilization; 7. Husbanded Resources). Contacts that we made while we had the ATE grant have been maintained through electronic connections, and we make use of an ATE-supported center's resources for our technical area (1. Wide Participation; 3. Resource Mobilization; 4. Knowledge and Skills/Training). When someone gets the opportunity to use the small amount of money we do have to receive professional development, s/he provides a seminar for the rest of the faculty (7. Husbanded Resources).

A COMPARISON OF THE ATE PROGRAM TO THE SUSTAINABILITY CHECKLIST BASED ON SURVEY/SITE DATA

In addition to providing the previous portrayal of a successfully sustained ATE project, we believe it is useful to determine how the different sustainability checklist elements are manifesting themselves in the ATE *projects*. As a shorthand technique, a rating of the degree to which the survey and site visit data from the Western Michigan University (WMU) evaluation project show that the ATE program is engaged in each element is suggested below. Each element was rated on a 1–4 basis: (1 = seldom evident, 2 = sometimes evident, 3 = often evident, 4 = almost always evident).

The WMU evaluation project has two major sources of data about the ATE program – a set of 13 site visit reports and 2 web-based surveys, all conducted in 2000 and 2001. Details of these studies are provided in the overview document (Chapter 1). The data were not gathered in a way that allows for causal analyses, nor were data available on the quality of the activities except in relation to the site visitors' expert opinions. Furthermore, data were obtained only from presently funded *projects*, so no direct evidence is available of what was or was not sustained. Approximately 20 items on the survey relate to sustainability. Eight of these are items that rate *project* growth, and the rest pertain to the different *project* activities such as collaboration or materials development. During the site visits, the site visitors were asked to comment on the potential for sustainability of the *project* overall, and additional comments about sustainability were made in the individual sections related to the ATE program activities. Comments related to sustainability in the site visit reports were marked, reread, and considered in conjunction with the survey data.

(1) Wide Participation and Clear, Shared Purpose – 3 (Often Evident)

This first element relates mostly to the collaborations required by the ATE program and the development of a shared vision within these groups of what each *project* is to accomplish. The ATE program appears to have wide participation with a variety of groups that have personal interest in different elements of the *projects* being sustained.In both years of the survey, a majority of the *projects* reported increasing or substantially increasing direct participation by other institutions and organizations. In each year of the survey, the median *project* lists at least 22 or more collaborative efforts, and each engages slightly more than 2 people per collaboration. The most prevalent collaborator reported by *projects* in both years of the survey is business/industry followed by educational institutions. When asked what factors affect the quality of the collaborative efforts, the PI responses in both years were categorized into 5 areas: (1) the quality and the enthusiasm of the people involved; (2) commitment/interest; (3) mutual benefit; (4) common purpose/vision; and (5) communication including clear expectations.

During the visits, all sites mentioned advisory committees of some sort comprised of business/industry representatives. The use of advisory committees for workforce programs is generally required at community colleges, often along with very structured program approval processes. These requirements contribute to the sustainability of the collaborations and therefore to the tie of educational programs to business/industry needs. Several *projects* are supported by business and industry because of their needs for employees and their lack of preparedness to provide introductory level training. The ATE *projects* are seen as providing good service at a favorable cost; as a result, these *projects* receive business/industry support and endorsement.

Projects also collaborate with professional organizations or work to build them up. These collaborations can provide sustainability of *project* activities in terms of prestige and name recognition. In one site visited, the ATE group became a quasi-national communication hub, and it is likely that the communication function will continue after funding ceases. Several of the professional organizations are business/industry related rather than education related. This contributes further to connections with business/industry. In two visited sites, the ATE *project* personnel had positions of national prominence in the business/industry professional organization. This clearly enhances awareness of the efforts of community colleges and their suitability for meeting business/industry needs. Connections with these professional organizations also provide additional venues for future funding as it becomes necessary. Having a national presence also helps local programs and their staff members believe what is being done is important and worth continued funding.

The survey and site visit data show some potential barriers to sustained collaboration in terms of wide participation and shared purpose. Many collaborations are based on the energy and commitment of one person – either from business/industry or from the *project*. Having these dynamic leaders is wonderful as long as the leaders stay at their institutions and remain interested in their programs; but if not, programs often disappear when their champions do. Additionally, because many of the businesses/industries are strictly local, they are reluctant to support efforts with a national focus. Interviews during the site visits reveal that the purposes of the ATE *project* activities are not always shared nor understood in the same way by all involved in the collaborations. On the survey, the projects reported communication and coordination as barriers.

(2) Abundant Information Available and Used to Improve the Program and Reward Effort – 2 (Sometimes Evident)

This element is directly relevant to the management and evaluation of the *projects*. Many different types of information are necessary for something to be sustainable, and this information must be used effectively to provide accountability information and to reward effort. First and foremost, the information needs to be used to determine if the activity is worth sustaining. Also, if mechanisms are not in place to continuously provide the information necessary for incremental change and if that information is not acted upon, the *project* will remain in the past and will become obsolete. The notion of reward is a subset of this element, since it is a special use of information that also provides incentives to continue a particular behavior.

The site visits and survey data showed that there is only some carefully collected and targeted information available. What is available is not always used nor is it necessarily related to program improvement. It is often information for information's sake. At the end of this paper, specific recommendations for data to be collected to track progress toward sustainability are provided.

In each year of the survey, more than 80% reported use of a *project* evaluator. However, based on the site visits, it appears that only modest data collection for evaluative and accountability purposes is undertaken (e.g. collection of number of students enrolled, number of students completing programs, number of students that gain credit for articulated courses, number of students in mentoring programs, and number of students that go on to college and their majors). Without this critical information, *projects* are most likely "operating in the dark," not knowing how to improve activities and programs and/or reward effort, not knowing if the goals for their activities are being attained, or unable to track their progress toward sustainability.

Directly related to this is the information collected for materials development. Although business/industry people are consistently involved at the beginning of materials development, use of other subject and pedagogical experts is less common. Pilot testing and revising materials by the creators based on informal input from classrooms and students are also very common. There is much less collection and use of more formal data on the effect of materials on student outcomes or on effects in diverse settings.

In terms of reward, there appears to be very little acknowledgement of effort for community college faculty to improve themselves through the ATE program. The work on the ATE program often seems over and above regular job responsibilities. Although it is seen as valuable, work on the ATE program rarely fits into the usual reward structure. There is very little money available for professional development, and it is not rewarded in the standard system. For example, one community college visited doesn't allow release time for its faculty members, and another doesn't offer ATE professional development for credit, which means it doesn't count on the merit pay schedule. This is particularly disheartening because, as one *project* reported, it is difficult for the faculty to adjust to new project-based approaches to learning; and they need the type of assistance that professional development could provide. Furthermore, instructors at community colleges are often paid less than similarly qualified people in business/industry, making recruitment and retention of these instructors a major challenge, especially after they have participated in extensive professional development. Perhaps if the value of professional development was clearly demonstrated, the community colleges would find more money to support this activity.

(3) Abundant and Needed Resources, Resource Mobilization – 4 (Almost Always Evident)

This element highlights the need of any *project* to have the necessary fiscal, material, personnel, and emotional resources available. For ATE, this involves the support of business/industry partners, the home institutions, other collaborators, and peers. Necessary resources also include individuals to teach courses, develop materials, program computers, handle administrative details, etc., and materials required for individuals to carry out their tasks. A critical resource, of course, is the money to pay for the people, time, and materials.

The site visits and surveys show that the ATE *projects* are making quite good use of existing resources, bringing all available resources to bear on the activities, and using what they have to leverage more resources. However, this does not mean that the *projects* have all the resources they need. This seems to be particularly relevant in terms of faculty hires and salary and in the potential for continually upgrading high cost technical equipment.

The survey data show that for every dollar provided by NSF for the duration of projects'/centers' grant periods, the *projects* reported increasing their working resources for the ATE program by 50 cents in 2000 and by 80 cents in 2001. *Projects* also reported receiving more than $12,000,000 each year in direct contributions of money and $16,000,000 of in-kind support in addition to the NSF funding. Connections across peer institutions also provide monetary support. The *projects* reported themselves as increasing their receipt of financial support from other organizations and income from and use of developed products. On the other hand, they reported lack of time, money, or other resources as barriers.

Several visited sites reported sister institutions adopting their programs, which would save them the developmental costs. At other visited sites, an institution is designated to provide a given type of educational program for the whole system. A variation of this at other sites has institutions sharing the cost and effort of a technical program by offering different modules at different sites. Either variation makes a program viable though a system that would not be so at a single institution. Barriers to system viability of programs do occur when there is competition between the institutions or varying quality of offerings or special student needs. Additionally, in order to obtain support across a network of peer institutions, a *project* might have to broaden its original work scope to something such as high tech education rather than its original focus on education for a specific technical skill.

Another element of resources is materials for quality programs. Materials are resources in and of themselves and also secondarily because they can be sold to produce revenue. In each year of the survey, more than 1,000 of these materials were reported in use at least locally. If one presumed all materials developed were used at least on a local basis, then in each year, at least 35% of this total was used at sites other than the *projects* and 11% were commercially published.

Sustainability of materials is less related to their development per se as to their continued updating and availability. Only some of the *projects* visited have plans for what will happen to their materials once the grant stops; most do not. Only one appears to have a very clear process in mind, although it is a project that is solely focused on materials development. Two sites visited have business plans that are designed to make them self-sufficient; and the materials provision is within those plans, since selling materials is one way to attain income. Difficult issues such as who will update the materials, Web listings, and links to materials if a project/center is not sustained do not appear to have been considered by the visited sites.

Directly related to the issue of sustainability of materials development is the connection of materials to curricular program requirements. Materials that are required by programs of study or have some clearly defined marketability are more likely to be sustained. For example, one site shared its concern that its program could be negatively impacted as a result of the core tech area of its ATE program being dropped from its state's high-stakes high school proficiency test and most likely the high school curriculum. The site visits showed that the materials developed are not always linked to programmatic requirements or even market demand. In terms of market demand, most of the materials produced are community college instructional materials for teachers. Unfortunately, there is low market demand for these since few community colleges have the funds necessary to purchase them. Student materials are more likely to have a market.

(4) Knowledge and Skills/Training – 3 (Often Evident)

Knowledge and skills training within the ATE program can be applied to at least four distinct groups – community college students, high school students, high school teachers, and community college instructors. This element, because of the last two groups, also directly relates to professional development and points out that specific knowledge and skills are necessary to achieve the goals of the *projects*.

For community college students, the numbers of students in a program are strong indicators of a program's sustainability. Because of the different emphases of the *projects*, these numbers varied substantially for the visited sites. Some had very high numbers of students (e.g. 700) while others had small numbers (e.g. 7–10). Generally, the numbers of students involved at the visited sites were modest (i.e. 10–30 students).

Enrollment numbers reported on the surveys were more encouraging. For *projects* as a whole, average enrollments reported for the past 12 months more than doubled in the secondary and associate degree level courses (244 vs. 700 in 2001 for secondary; 915 vs. 2,300 in 2001 for associate). However, these increases appeared to be due largely to a few institutions rather than an across-the-board increase. For example, at the secondary level, two *projects* reporting for the first time in 2001 cited enrollments on the high end of the range (2,000 and 5,000). Similarly, one center reported enrollment in 2001 of around 70,000 in its associate level programs. The *projects* also reported themselves as increasing for numbers of students enrolled, students placed in technical jobs, and students graduating or completing programs.

When viewed from the perspective of a specified program conducted during the last 12 months, *projects* reported that their average enrollment rose substantially (94–160 students in 2001 at the associate degree level). In these specified programs, the average number of program completers also increased from 43 to 58 in 2001 at the same degree level. When examined from a per course perspective from the survey data, the average number of students enrolled in a course in a specified program increased from 7 to 11 in 2001 at the associate degree level. However, based on our site visits, at many colleges the break-even point for providing instruction is 15–20 students per course.

Despite these strong indicators of potential sustainability for programs overall, the sustainability of the new mentoring or internship opportunities funded by ATE is more questionable. These opportunities require additional resources and are likely to decrease or stop without continued external funding or explicit program requirements for these experiences.

Knowledge and skill training of high school students takes two forms: directly to high school students or indirectly through the education of their teachers. Several visited sites provide both types of educational opportunities. Students are often trained directly, although these approaches are generally more interest-generating opportunities such as science camps or opportunities to learn about exciting scientific advances. Two visited sites also report working with high schools to provide general upgrading of students' academic backgrounds. Another approach used at one visited site is working with high school counselors and parents to enhance the image of a two-year degree in comparison to a 4-year one. One project has both community college and high school students working at its partner institutions at the very jobs they are training to obtain – real on-the-job training. The service this provides to the participating institutions increases hope for sustaining the activity.

Presently, there is a great deal of professional development in the ATE program. The participants find these opportunities valuable, but little is known about the implementation of the education received. Also, although the training is rightfully focused on content upgrading, other types of knowledge (e.g. instructional materials development) are less apparent. A confounding issue is one also mentioned above under resources: it appears unlikely that the level of professional development presently supported will be sustained. In other words, there are few processes in place for providing the continuous upgrading and cutting-edge knowledge necessary to maintain excellence in technology. Despite this, it is likely that the upgrading already provided will improve instruction.

ATE-provided professional development opportunities are well attended and received. In both years of the survey, conferences, workshops, and in-service courses were the most popular formats. Course offerings were well attended with a median of around 20 individuals for projects and about 130 for centers. Most participants were from associate degree granting institutions. Regarding how full their professional development opportunities were in both years, more than 75% of the projects reported they were at least at 75% capacity, and more than 90% of centers reported this level of capacity. On the 2001 survey, the *projects* reported themselves as increasing in the number of and participation in professional development opportunities.

High school teacher education programs are often associated with the development of articulation agreements between high schools and community colleges, and the formal agreements will help to sustain the associations. Regardless of the development of formal agreements, the teacher education programs update the high school teachers' technical knowledge and help guarantee that the high school courses will be consistent with the courses offered at the community colleges. The provision of educational opportunities for high school teachers and students is likely to be sustained only if these encourage enrollments and or enhance the public image of the sponsoring institution thereby increasing its chances for external support.

Despite these successes and as mentioned in item no. 2, there are few processes in place for providing continuous upgrading and cutting edge knowledge other than the ATE program. The *projects* also reported the ability to attract/keep faculty and other critical staff members as a barrier as well as faculty having difficulty adapting to the changes needed for the new programs.

An issue related to professional development of existing faculty is the use of adjunct faculty. Many of the visited sites reported very small numbers of regular faculty and large numbers of adjunct faculty. This is a positive development if the adjuncts are providing knowledge based on recent contact with cutting-edge business/industry techniques. On the other hand, this does not bode well for sustainability because, if money gets tight, adjuncts are most likely the first to go. Additionally, if a program is staffed almost completely by adjuncts, there is no ongoing supportive voice in the regular processes of the institution and hence little hope for institutionalization of the program (see item no. 6).

(5) Decision Making/Distributed Power – 4 (Almost Always Evident)

This element relates to the consideration of shared power because, in that way, more people are responsible for the success of the *project*, and more people feel ownership. This also allows a *project* to continue essentially undamaged in the event that key personnel leave the project. The ATE *projects* seem to operate fairly distinctly from their individual schools and from NSF. They also appear to share the decision making related to their *projects* through their collaborations. On the other hand, the success of many *projects* appears to be centered in one dynamic PI.

(6) Coordination with Current Initiatives, Administrative Support – 4 (Almost Always Evident)

This element presents the issue of institutionalization of a *project*. To be institutionalized, a *project* must become part of the fabric of the organization in which it is embedded. It must fit with and complement the other institutional initiatives and goals, and it must also meet its needs utilizing existing institutional processes. To guarantee that it can continue, a program needs to be formally approved, have its own permanent faculty, and a continuing budget line. Even smaller components such as courses need approval and support.

Chances for institutionalization are highest if a similar type of program existed before the grant was ever obtained. In other words, sustaining improved programs is easier than sustaining newly created programs. A preexisting program is already institutionalized; therefore, only the new elements have to be incorporated into the system. Incorporation of new elements is much more likely if these elements can be provided by existing, full-time faculty or if new, but full-time, faculty lines have been authorized. This was actually the case at several of the site visit locations. The visited NSF *projects* expanded or improved on existing programs, and it seemed highly likely that these programs would continue after the grant expired. These improved programs were delivered by existing faculty or new, full-time faculty members permanently hired to deliver the program.

New programs can also become institutionalized. This is most likely to occur if there is strong external support for the program, such as industry demand. It is also helpful if the principal investigator promoting the program is a well-respected faculty member with institutional power, as is often the case in ATE *projects*. Administrative support is also crucial. The visited sites report high levels of administrative support for both new and improved programs. The administrators are creative in diverting resources, and they are often the driving forces behind the programs. The administrators report viewing the ATE programs as models of what could be done in other areas and see the ATE programs as showing other faculty that they could get funding as well. However, administrator support for programs is tempered. They are supportive only as long as the programs provide incentive or revenue. One administrator stated that he is supportive as long as there is a continued increase in enrollment. Another said he could only support the purchase of the high tech equipment for a few years and then it would have to stop because the prices were prohibitive even if enrollments were high. On the survey, the *projects* reported administrative support as a barrier.

Another condition that enhances sustainability is the development of spin-off programs or other opportunities that feed on and nurture the first effort. At one visited site, a second program was spun off from the one supported by NSF. In three other cases, large development projects, such as technology centers or industrial parks, provided impetus to guarantee the sustainability of the programs.

Capitalizing on other funding also enhances the potential for sustainability. Several *projects* are working with other grant programs such as tech prep (a program to help students move from school into technical jobs) to help support efforts related to their ATE *projects*. Tech prep is particularly relevant, since it may enhance the image of technology overall and provide technological literacy to all students. Sometimes, specific grant funds were made available for the training programs of technology education teachers. As mentioned under item no. 1, direct ties with businesses/industries that need to hire students are also supportive of sustainability. This is a complex issue, however, because many technical businesses are small and do not need large numbers of employees (e.g. biotech) or they need workers in different parts of the country, and students at community colleges often do not want to move.

(7) Use of Promotion and Marketing/Husbanded Resources – 2 (Sometimes Evident)

This seventh element relates to sustainability in terms of the *project* as a business. In order to continue, a *project* must be believed to be necessary, and marketing is one of the best ways to accomplish this. To be sustained, the ATE *projects* need to convince their various constituencies that they are worth continuing. This approach also assumes some sort of outcome, process, or product that is to be marketed or provide value. For ATE, these are generally related to instructional materials or professional development opportunities. Additionally, the husbanding or targeting of resources allows a *project* to be more flexible, hold out in lean times, and take advantage of unexpected opportunities, all contributing to sustainability.

In keeping with NSF's evolving interest in sustainability, the centers and newer projects show more evidence of meeting business goals than the earlier funded *projects*. Many projects have Web sites, and some are quite complex and entrepreneurial. The PI meetings also promote this business goals attainment approach. However, it is a relatively new approach for NSF, and many of the *projects* do not see it as a reasonable goal.

The commercial textbook and instructional materials companies are often more highly funded than the ATE projects and are therefore more competitive. Furthermore, the locally developed ATE materials are often specifically designed to fit with a particular site and designed for teacher and not student use. People at other sites are reluctant to put in the effort involved to adapt or modify the materials for their use.

A thorny issue is related to marketing materials or professional development in order to provide sustainability. ATE *projects* need to reconcile the fact that they are supposed to be providing the best materials and education available to their fields with the reality that they may have to sell materials or professional development if they are to sustain themselves. Just how entrepreneurial should projects be? Should they employ marketing experts? Should they withhold valuable materials or education from colleagues because they can't afford them? One visited site has grappled with this issue by marketing its services at different rates based on ability to pay, but this solution presents difficulties as well. These issues are inextricably intertwined with dissemination issues.

Another aspect of marketing and husbanding resources that affects the sustainability of a program is its ability to train people for a variety of situations. This approach tends to increase numbers of students in courses because the courses offer material that is valuable for students pursuing different areas of expertise. It can enhance employability of the graduates in the face of changing job markets and thereby increase enrollment because more students see the program as a pathway to employment. It also allows the educational institution the flexibility to change direction. Several visited sites mentioned that they are providing a broad-based set of experiences for their students, they "were not putting all their eggs in one basket" (e.g. all into network administration training), they are involving instructors from a whole division, and/or they are using several different platforms (e.g. several types of network software).

CONCLUSIONS AND RECOMMENDATIONS

As can be seen from the Fig. 1, overall, the ATE *projects* appear to be making progress toward sustaining themselves in some form after the NSF monies are no longer available. Embedded in these findings is the importance of achieving and documenting concrete steps toward accountability. Formal certification or approval of these steps from advisory committees, administrators, professional organizations, and others is important. Examples of approvals include creation of new permanent positions, official course or program authorization, or a state's determination to include technician-based content in its curriculum standards or assessments.

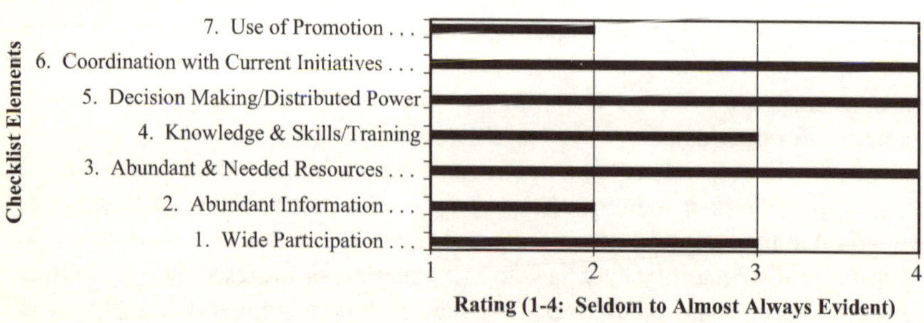

Fig. 1. ATE and the Sustainability Checklist.

Based on data from the site visits and surveys, there is strong evidence indicating that the ATE program manifests 5 of the 7 elements necessary for successful sustainability from our literature-based sustainability checklist. Two elements that we believe need monitoring and improvement are: (1) the availability of abundant information and its use to improve the program and reward effort (checklist item 2); and (2) the use of promotion and marketing/husbanded resources (checklist item 7), since there is only some evidence indicating that these elements are present in ATE.

A specific set of recommendations we believe will assist the ATE program to optimize sustainability for its *projects* is detailed below. These recommendations are intended to ensure sustainability, however it is defined, by increasing the likelihood that the 7 elements necessary for successful sustainability of any program will manifest themselves in each ATE *project*. These recommendations also address the areas of concern identified for the ATE *projects* under each sustainability checklist item, especially items 2 and 7.

(1) NSF Should Clarify Its Position on Sustainability for the ATE Program

As stated at the beginning of this paper, there is very little information in NSF publications pertaining to sustainability. It appears to be an evolving issue. As a result, ATE *projects* may not be aware of NSF's interests in this area. Even if *projects* are aware of NSF's interest in sustainability, they are unlikely to have clear ideas for operationalizing it or for determining what should be sustained. NSF may want to include the PIs in developing a definition of sustainability as it applies to the ATE program.

(2) If Sustainability is to be a Major Goal of the ATE Program, NSF Should Consider How Best to Help Projects Achieve It

From our work on the evaluation project, we have observed the increased synergy that occurs when *projects* share programmatic information and strategies. There are already two vehicles for creating this synergy for sustainability – NSF's Web site (http://www.nsf.gov) (or the evaluation project Web site [http://www.ate.wmich.edu]) and the PI annual meeting. NSF may want to include sessions on sustainability at the PI meetings. NSF may also want to provide advice on how to consider and achieve sustainability to *project* national visiting committee members. Having a Web seminar for the PIs and/or creating an electronic mailing list (or chat room) are other options. NSF may also want to consider providing technical assistance (e.g. dollars for a consultant for marketing/promotion, technical assistance for data collection methods) and other interventions. Ultimately, NSF may want to add explicit consideration of sustainability to the program announcement and a rating system to help reviewers assess a potential grantee's sustainability plan.

Detailed suggestions for a sustainability plan are included in recommendation no. 4.

(3) More Attention Should be Given to Data Collection and Use to Identify Project Components that Should be Sustained, Learn How to Improve Components, Provide Information Upon which to Base Rewards, and Convince Others of the Worth of the Components

As pointed out under item 2 of the sustainability checklist (Abundant Information Available and Used to Improve the Program and Reward Effort), many *projects* are struggling with measuring their progress toward goals and providing data to inform key stakeholders (e.g. NSF, administrators, business/industry partners) because they do not have effective strategies and methods in place to collect critical indicators of progress. As a result, these *projects*, because they do not really know how they're performing now, can't really improve or reward performance nor effectively plan for the future. Data should be gathered to help determine what should be sustained and how. The list of issues discussed in this report might help identify types of data to include (e.g. items on the sustainability checklist).

The NSF FastLane report and the annual evaluation survey are two existing mechanisms for assisting with the collection of key indicator data. The old adage in many business/industry settings is that "what gets measured, gets attention." We have observed that because of these two mechanisms, *projects* are now starting to attend to specific kinds of data (e.g. number of collaborations and materials developed; enrollment and completion figures). The collection and use of critical information for *project* accountability and decision making could be further expanded by developing and disseminating a set of key indicators for meeting ATE program goals and accompanying means to collect data for these indicators. Assessment of progress toward sustainability could also be an indicator (How this assessment would be accomplished is described in Recommendation 5 below). These key indicators would also need to be emphasized on the FastLane report and the annual evaluation survey.

(4) ATE Projects Should Consider Integrating Sustainability Strategies Into Their Work from the Outset

We believe that the detail we are suggesting for inclusion in the sustainability plan described below ensures that the critical elements necessary for sustainability are given attention from the beginning of a *project*, thus strengthening the probability of achieving this goal. The recommendation assumes that the *project* is worth sustaining. This recommendation is also aimed at the concerns raised in regard to the varying degrees of administrative support, promotion/marketing, and institutionalization.

We suggest the development of a multiyear plan, perhaps six years to mirror two NSF grant cycles, that is based on an "exit strategy" for NSF's funding (i.e. what are the strategies for replacing NSF monies and other support over time). Elements of this plan, including implementation strategies and schedules of key milestones (e.g. written commitment from administration), could include the following:

- The ongoing vision and goals (i.e. where a *project* sees itself and its partners in six years).
- Methods and timetables for collecting data critical for determining quality, accountability, and decision making and means for sharing and using this information with key stakeholders.
- Identification of and strategies for obtaining additional funding, revenue sources (e.g. product income), and other support (e.g. dollars for training, release time for faculty) outside of and/or beyond the time of the NSF grant.
- Descriptions of collaborations/partnerships and what the contributions of these partnerships will be in concrete terms (e.g. specific goals, in-kind services, dollars, dissemination of materials – written commitments and for what time period).
- A depth chart (i.e. list of individuals who could step in when key personnel and partnership changes occur), including contingencies for critical personnel and partnership changes (e.g. PI change, key business/industry partner struggling through an economic downturn, loss of another key funder).
- A description of the strategies for incorporating the *project* within the institution (e.g. approval of a tech program by a specified date, plan for written commitments from administration over time).
- A promotion and marketing plan that outlines the various means (e.g. Web, conferences, publications, professional organizations, trade shows) to be used to raise awareness and acceptance of a *project* and to update and disseminate its products (e.g. materials).

(5) Assessment of Progress Toward Sustainability Could Occur at Least Annually

As the *projects* are implementing their sustainability plans, they could assess their progress toward sustainability at least annually as a reality check. This self-assessment would involve using the sustainability checklist described in this paper with key stakeholders and collecting artifacts as described in Recommendation 4 that provide evidence that progress toward sustainability is occurring (e.g. agreements with administration for providing a permanent faculty member once the ATE funding is done). We also suggest that NSF add a question to the FastLane report asking for the general results of this self-assessment. This not only ensures that this self-assessment will be done, but also informs NSF on an annual basis. Adding a question or two to the annual evaluation survey so that results would be aggregated across *projects* is also another suggested method. As many *projects* have indicated they focus on the National Site Visiting Committees' recommendations, NSF may also want to have these committees use the sustainability checklist and collect artifacts as part of its evaluative process or at least review and comment on the site's self-assessment.

NOTE

1. A large portion of the discussion about sustainability in the literature focuses on the sustainability of the environment or maintaining our society within environmental limits. Another segment addresses economic sustainability.

REFERENCES

Cropper, S. (1996). Collaborative working and the issue of sustainability. In: D. Huxham (Ed.), *Creating Collaborative Advantage* (pp. 80–100). London: Sage.

Flower, J. (1996). The five fundamentals of dealing with change. *Physician Executive*, *22*(8), 37.

Gabelnick, F., MacGregor, J., Matthews, R., & Smith, B. (1990). Making learning communities work: Issues of implementation and sustainability. *New Directions for Teaching and Learning*, *41*, 39–51.

Howard, D. C., & Howard, P. A. (2000). Towards sustainability of human services: Assessing community self-determination and self-reliance. *Canadian Journal of Program Evaluation*, *15*(1), 25–40.

Schneider, B., Brief, A., & Guzzo, R. (1996). Creating a climate and culture for sustainable organizational change. *Organizational Dynamics*, *24*(4), 7–19.

Schwartz, A. (1994). Eight guidelines for managing change. *Supervisory Management*, *39*(7), 3.

THE ROLE OF ADVISORY COMMITTEES

Wayne Welch and Arlen R. Gullickson

ABSTRACT

The findings from the evaluation of the Advanced Technological Education (ATE) program are described in a set of documents assembled by The Evaluation Center at Western Michigan University. The documents were examined to determine what information was available on the use of advisory committees. The purpose of the analysis was to improve the effectiveness of ATE committees.

Limited information on the use of advisory committees was found in the data sources examined. The information that was available dealt primarily with centers and larger projects. Virtually no information was available on the use of advisory committees in the smaller ATE projects. These comprise 56% of the total number.

Generally, the reactions by ATE grantees toward committees was positive. They are commonly used in community colleges and in vocational/technical education programs. Most of those who responded found them useful for collaboration and identifying work force needs. However, a bias toward larger projects and the limited number of comments available precludes generalizations to the total population.

Because little is known about how ATE advisory committees function, their organization, and usefulness, especially among most of the grants, additional research is recommended. A set of research questions and a research design

Evaluating the Upgrading of Technical Courses at Two-Year Colleges:
NSF'S Advanced Technological Education Program
Advances in Program Evaluation, Volume 9, 185–199
Published by Elsevier Ltd.
ISSN: 1474-7863/doi:10.1016/S1474-7863(04)09009-X

are suggested. Some preliminary ideas on the elements of effective advisory committees are presented.

FORWARD

The goal of the Advanced Technological Education (ATE) program is the improvement of technological education at the undergraduate and secondary school levels. The program was authorized by Congress and is funded by the National Science Foundation (NSF). Since its inception in 1994, about 500 grants have been made, primarily to 2-year colleges. About half of those ($n = 220$) were active in 2003.

A contract was awarded to Western Michigan University (WMU) in 1999 to evaluate the ATE program. Part of the work of the WMU evaluation project is to examine several key issues related to the implementation of the program. An issue is defined here as a topic or situation that requires further consideration to make a decision or reach a conclusion. Five issues addressed by the ATE evaluation come from the core program objectives mandated by Congress. These issues include dissemination, materials development, collaboration, professional development, and program improvement.

Four other issues evolved from discussions and concerns expressed by NSF program officers and the ATE evaluation staff. They are recruitment and retention, sustainability, evaluation, and advisory committees.

These issues are explored in a set of nine papers, called Issue Papers. The papers were selected as a way to interpret and report the findings from the WMU evaluation studies. They are based upon information contained in two status reports[1] and the reports of site visits to a sample of 13 projects and centers. Additional context was obtained from the program guidelines, NSF staff, literature reviews, and the involvement of the authors in various aspects of the ATE program. This included, for example, serving as a project evaluator or as a member of a site visit team.

The papers were written by different authors, each with their unique background and experience. As such, they vary in tone, content, length, and how much judgment and recommendations they contain. However, all are designed to provide information to help make decisions or reach conclusions about the issues in the ATE context.

The major audience for the Issue Papers is the ATE staff at the National Science Foundation. Secondary audiences are the staff of the current ATE grants and those who might be thinking about submitting a proposal to the ATE program. Science, mathematics, and technology educators may also find these papers useful as they consider ways in which to improve their own technology education programs.

INTRODUCTION

Advisory – having or exercising power to advise. *Merriam-Webster Dictionary*
Committee – a body of persons delegated to consider, investigate, take action on, or report on some matter. *Webster's New Collegiate Dictionary.*

> "I have found the best way to give *advice* to someone[2] is to find out what they want and then advise them to do it." *Harry S. Truman*
> "A *committee* is a group that keeps minutes and loses hours." *Milton Berle*

There is a wide range of feelings about committees in general, and advisory committees in particular. Nevertheless, they are a common element in the educational, political, and social fabric of our lives. Something in our being tells us that when a problem is encountered or work is to be done, a committee is an effective way to proceed. But many others consider them a waste of time.

Harry Chapman, writing to his colleagues in the medical profession, said:

> Having served on various committees, I have drawn up a list of rules: Never arrive on time; this stamps you as a beginner. Don't say anything until the meeting is half over; this stamps you as being wise. Be as vague as possible; this avoids irritating the others. When in doubt, suggest that a subcommittee be appointed. Be the first to move for adjournment, this will make you popular, it's what everyone is waiting for.[3]

Despite being the targets of many critics, committees survive and proliferate. In recent years, advisory committees have become an obligatory requirement for proposal writers, grantees, and funding agencies. The reasons for this are seldom examined. It seems taken for granted that a committee will increase the likelihood of success or at least give the appearance of doing so.

The Advanced Technological Education (ATE) program funded by the National Science Foundation (NSF) is no exception to this trend. The projects funded by this program are encouraged to use advisory or steering committees. In addition, centers and larger projects are required to have a national advisory committee review their work annually. The national program is reviewed every three years by a committee. ATE is also being evaluated by the Western Michigan group which, in turn, has its own advisory committee. It would appear that a great deal of advice is being sought from several different groups of advisors.

In this chapter, ATE advisory committees and their relatives[4] are examined using the information obtained from the Western Michigan University (WMU) evaluation. The purpose of the chapter is to consider that information to help determine whether advisory committees are useful and, if so, how they might be improved.

USE OF ADVISORY COMMITTEES

The first step in the process was to examine the available data sources for information they might contain. The sources were ATE program guidelines, the reports of the site visits conducted by the WMU group, two status reports published by WMU, and the results of a Web-based survey. Each information source was searched for references to "advisory committees" or similar words. In addition, a limited search of the education literature was conducted.

ATE Program Guidelines

Two ATE program announcements and guidelines, published by NSF, were examined to see what suggestions were given to proposal writers. The 1998 guidelines did not mention advisory committees, but the 2000 guidelines contained specific information on their use. Larger projects[5] and centers are now required to have National Visiting Committees (NVC). A brochure was published by the Foundation explaining the purpose and procedures of these committees.[6] According to the brochure:

> Each National Visiting Committee normally consists of six to eight (although some may have as many as 12) national leaders in the fields of science, mathematics, engineering, technology, evaluation and education. To assist projects in achieving their full potential, NVCs: (1) advise the project leadership based on the collective expertise and perspectives of the NVC membership; (2) assess the project's progress and plans, and report their findings to NSF and project leadership; and (3) provide assistance in the specific areas as identified by the NVC.

Typically, the National Visiting Committees of large ATE projects include four to six members, while centers have eight to ten people on their committees. NSF program officers work with grantees to work out details of membership and budget. The program officers are ex officio members of the National Visiting Committees. In a reply to e-mail queries, the program officers said they do not go to all NVC meetings. For centers, they try to go to the first one and to the one that occurs before their renewal request after three years. Some meetings have been "attended" via the telephone or two-way video.

One NSF staff member estimated that about 80% of the larger projects have some kind of advisory committee, usually less formal than the NVCs. Some of these committee meetings have been attended by NSF staff members. In other cases, former NSF program officers attend meetings when a current program officer cannot attend.

The guidelines are silent on the use of advisory committees for the smaller projects, that is, those with funding less than $500,000. According to figures provided by the WMU group, these represent about 56% of all ATE grants funded.

Results from the Site Visits

The Western Michigan team arranged for a series of visits to 13 sites selected to represent the range of grants funded by the ATE program. A list was provided to the WMU team of 2 dozen projects and centers that NSF program officers thought would be willing to participate in a site visit. The team selected 13 from that list to represent the variety of grants funded by the ATE program. The selection criteria included size of grant, geographic location, technology focus, and type of institution receiving the award, e.g. 2-year college, 4-year college. The sample included 3 centers, 5 large projects (funded for $500,000 or more), and 5 smaller projects.

The site visits were carried out by two or three people contracted by Western Michigan. They included members of the ATE evaluation staff, members of the Evaluation Advisory Panel for the WMU project, and external evaluators who have worked with the larger projects and centers for the past several years. The visits were of one or two days' duration. A report was prepared and submitted to the grant staff for review. The revised reports were used in the preparation of the issue papers.

The 13 reports were searched for references to advisory committees or similar terms. Seven reports mentioned either advisory or national visiting committees. The search terms were found 26 times and appeared in 17 paragraphs. All of the references found were in the site reports of either centers or larger projects. There was no mention of committees in the reports from the 5 smaller projects.

Five reports stated that advisory committees were used for collaboration with constituencies, primarily industry. An example is "the state and the college use industry-based advisory committees to review curriculum development to ensure relevance to the workplace." According to the reports, this collaboration helped the colleges identify job skills, monitor industry trends, and provide internship sites and scholarships. The industrial partners on the committees also donated laboratory equipment and provided facilities for on-site instruction.

Also, five reports (not necessarily the same five as above) mentioned the general use of advisory committees by community colleges. They are used in many fields of study to help the colleges respond to changing community needs. For example, 1 site visit report stated its community college had 600 people serving on its various advisory committees. Of course, listing people as committee members is one thing. Serving an active and useful role is another.

Four site reports mentioned National Visiting Committees (NVC) or, as they were called by some projects, National Advisory Boards. Thirteen comments about these committees or boards were found in these four reports. For example, "the project has an external evaluator who is doing a formative and summative evaluation. He provides information for NSF annual reports and to the national visiting committee."

Four comments were merely statements that members of the national committee were interviewed to obtain information for the site visit. Four others expressed favorable reactions toward the committees; for example, they provided good advice or positive support to the staff. In another comment, it mentioned only that an NVC exists. Another comment was that an NVC recommendation the project followed turned out not to be such a good idea. The final and thirteenth comment reported the NVC only repeated information given to it by the project's evaluator.

Because the above information was found only in the site reports to centers and larger projects, generalizations cannot be made to all ATE projects.

Status Reports

Status Report 1 prepared by the Western Michigan evaluation group mentioned committees five times. Part of the discussion was about the NVCs mentioned above. It described the opportunity these committees provide for interactions among the large projects and centers. Principal investigators serve on the NVCs of one another's projects. The report states that the program officers believe these interactions help them share ideas and establish a unity of purpose.

The report also mentioned a Committee of Visitors that reviews the national ATE program every three years. It is a small group of people. The first committee included a two-year college dean and faculty member, an industrial representative, and a person from another government agency that funds two-year college projects. Apparently, this information is used for program guidance and to provide justification of program success to the Foundation leaders and to Congress. No information was contained in the report about the effectiveness of this committee.

Status Report 2 mentioned the word "advisory" once. This was in a table that listed several purposes of collaborations by the type of organization. One purpose of collaboration is to seek advice, and the survey listed "advisory panel" as one example. The report provides figures for the percentage of responding grantees seeking advice from business/industry (79%), educational institutions (75%), public agencies (54%), and other (45%). This item was answered by slightly more than half the 113 grantees that received the survey.

Open-ended Survey Responses

The WMU evaluation project used a Web-based survey to gather information from ATE projects and centers. In the spring of 2000, it was sent to 113 respondents working with grants that were currently in operation. The group at Western Michigan University provided verbatim responses to the open-ended questions in the sections that pertained to Collaboration and Finalization. One example was, "Describe the nature and extent of your project's collaboration with other projects." Another was, "Describe up to three barriers to success that occurred in your project."

Eighty-four sites responded to the open-ended questions in the "Finalization" section. About one-fourth this number responded to the Collaboration sections. There were 128 pages of printout from the finalization questions and 32 pages of comments from the collaboration items.

The responses were searched for the word "committee." There were 25 occurrences, all but 4 from the Finalization questions. The sentences or phrases containing the search term were analyzed and placed into categories. These categories and the frequency of mention are presented below.

The committees are a way to involve women and minorities in our work. ($n = 6$)
They are effective for collaboration with other projects and centers. ($n = 5$)
Our project or center has an advisory committee. ($n = 5$)
They provide useful needs assessment information to give direction to our project. ($n = 5$)
The committees are helpful to us. ($n = 3$)
They praised our efforts. ($n = 1$).

In general, the people who mentioned committees were positive in their attitudes toward them. This is informative because studies of committee effectiveness are rare.[7] However, the number of respondents is small, and it would be risky to generalize to the total population. Furthermore, they were not asked specific questions about advisory committees but only mentioned them in answering other questions.

ERIC Literature Search Results

An examination of the documents contained in the Education Research Information Clearinghouse (ERIC) shows the common use of advisory committees by community colleges. ERIC contains more than 1 million abstracts of documents and journal articles. It includes documents on education research and practice from 1966 to 2001. A search of the database, using the key words "advisory committees," generated 3,471 references. Almost 10% (317) of these dealt with community colleges.

Advisory committees are also widely used in the vocational/technical education area. An ERIC search of "advisory committees" and "vocational and technical education" produced 953 documents, more than a fourth of the total. Many vocational and technical education programs are located in 2-year colleges. They rely upon ideas from business and industry to ensure that the college programs are meeting their constituents' needs. Advisory committees are established to obtain this information.

Few studies of committee effectiveness were found. Some evidence suggests they are not as helpful as many assume. Teitel reports the widespread existence of committees that exist on paper only.[8] Others have found that many programs and institutions do not effectively use their advisory committees for advice or support.[9]

Much of the interest in the use of advisory committees appears to come from the federal government and the programs they fund. In 1998, there were almost 1,000 federal advisory committees that were costing nearly $200 million yearly. Many federal agencies require advisory committees for the projects they fund. However, concerns have been raised about their effectiveness as well.[10]

SOME GENERAL CONCLUSIONS ABOUT ATE ADVISORY COMMITTEES

Although the information regarding advisory committees in the ATE context is limited, a few observations seem in order. These observations are based on the information gathered by the Western Michigan group; the literature review; and one author's experiences as a consultant to the project, the chair of its evaluation advisory panel, and a member of one of the site visit teams.

- Advisory committees are used by many community colleges and vocational/technical programs as part of their typical operating procedures. Using them as part of an external grant would be normal for them.
- The National Science Foundation expects centers and larger projects to have a national advisory committee. No information was found on use of advisory committees on the smaller projects. They represent about 56% of the total number of grants.
- Some respondents from the larger projects reported that committees are useful for assessing needs, especially in the industrial technology area. They also

reported that they are an effective way for collaboration, especially with traditionally underrepresented groups in the technological fields.
- Some projects having national visiting committees found them helpful for sharing information with other projects. However, at two sites the NVCs were not viewed as particularly helpful.
- Although there are many references to advisory committees in the education literature, there are few studies of their effectiveness.
- There have been no efforts to investigate ATE advisory committees directly, especially among smaller projects. The information that is available was obtained while investigating other elements of the ATE program, for example, collaboration or curriculum development. As a result, little is known about the actual functioning of the committees, e.g. membership selection, methods of operation, how to provide useful advice, contributions of members, cost of the committees, or their benefit to the program.

RECOMMENDATIONS

Because of the limited information that is available, the following recommendations are offered. First, there is a need to learn more about the ATE committees and the manner in which they function. Studies of committee effectiveness are rare, and none were found devoted to committees used by externally funded projects. Second, there is a need to inform those involved in ATE work about the tenets of effective committee work. This might require some kind of technical assistance, perhaps a videotape or a user-friendly handbook on advisory committees. These two lines of investigation are explored in more detail in the remainder of this chapter.

Learning More About ATE Advisory Committees

Answers to the questions listed below should make us more knowledgeable about advisory committee work. We expected that the original analysis would answer these questions. Because the data are insufficient to address them, they become the focus for additional study.

How many and which projects use some type of advisory group?
What is their purpose? Are the purposes appropriate, feasible to accomplish, useful?
How are they organized?
How are members chosen? Are they representative?
How do they function?

What do the committees do, and what products do they create? Are they relevant, of high quality, useful? How is advice transmitted from the committee to the project staff or other audiences? What is the impact of the committee?
Is it effective? What is the return on investment? What are the positive and negative side effects[11]?

These and other questions could be the focus of a research study conducted on a sample of current ATE projects. A random sample of 30 current projects, 24 projects and 6 centers, could be chosen. A set of research questions would be developed and administered to the grant staff and a sample of advisory committee members via a telephone interview. The research questions would be similar to those listed above. The information gathered from this research study would be used to provide guidance for ATE and other advisory committees in the future.

The details on the research questions and the information needed to answer these questions would be worked out following some preliminary investigation. It would be important to determine the concerns of NSF program officers, ATE projects, and the Western Michigan University Evaluation Center staff. While the initial data gathering is primarily descriptive, it may be necessary to conduct more intensive follow-up interviews to provide illustrative examples of effective practice and problems to avoid.

There is a body of research literature on advisory committees in general that could be gleaned for ideas. However, this literature is primarily devoted to long-term, volunteer advisory committees, not the limited term, paid advisors common in federally supported programs. In fact, no references in the ERIC search could be found for advisory committees that include paid consultants that serve a single project.

The most common documents in the system were about advisory committees for vocational technical programs, schools, and higher education programs. However, these were always volunteers who served on ongoing committees. Sometimes they found their advice was seldom followed, at least in a statewide study conducted in South Carolina.[12] Members serving on the mandated advisory committees for the state's vocational education programs were surveyed about their experiences. Seventy percent of the respondents reported their advisory group had no impact on the programs they served. Although the study had a response rate of only 41%, the findings are troubling.

What can we say to project directors who are asked to contact a half dozen experts to serve on an advisory committee that meets once a year for three years? Right now, not much! In fact, we cannot give them much assurance that the committee's work will be worthwhile except for the appearance of competence that a group of experts provides. Perhaps that is enough to justify their existence. Part of the proposed research would address that issue. (See the postscript at the end of this chapter for the follow-up research funded by the National Science Foundation.)

Some Ideas About Effective Advisory Committees

It is beyond the scope of the present chapter to provide a detailed set of guidelines for implementing effective advisory committees. However, a review of the literature was conducted to identify some important components of the committee process. Most of the information came from "this is how we did it" documents found in the literature. Other ideas grew out of the personal experiences of the authors. These notions are presented as first steps in defining a framework for thinking about improving committee effectiveness.

A noted physicist at Harvard University once said, "A committee is only as effective as its chair." When forming a committee, be sure a chair is selected who is effective at running meetings. A key goal for the chair is to keep the meeting on task, a constant threat when you have a committee of experts. Despite the claims by Harry Chapman (see p. 3), most people have a great deal to say before the meeting is half over! And many forget that an advisory committee meeting should be a meeting focused on a specific topic.

A former executive vice president of a large pharmaceutical company once said, "A meeting should not be held unless minutes are kept." He was concerned about the large number of meetings he had attended that turned out to be little more than an excuse for people to get away from their desks for some social interaction. He believed that minutes served as the record of accomplishment. This business executive also claimed that most meetings were much longer than necessary because of the constant straying from the topic.

Another seldom addressed element of effective meetings is proper facilities. How many of us have attended daylong meetings sitting on hard chairs in crowded rooms with the temperature too hot or the air conditioners too noisy? This situation is not conducive to productive meetings. However, many committee organizers seldom seem to consider it.

As one considers these rather practical considerations, they might be explained if meetings were viewed merely as a formality with little concern about accomplishment. A recent advisory committee meeting is a case in point. The federal contract called for the committee to meet during the first quarter, a "deliverable" as federal agencies are fond of requiring. However, final funding was received only a month before the end of the quarter. Several proposed committee members were not available on such short notice. Undaunted, the project director went ahead although only two committee members could attend! One wonders what a two-person committee should be called? A brace? However, the deliverable was achieved and dutifully reported to the funding agency. The project director considered the meeting useful, but the deliverable requirement seemed to drive the meeting more than a real desire for advice.

Other canons for committee meetings are often overlooked: for example, taking breaks but making sure that 10-minute breaks last only 10 minutes. Do not serve lunch while trying to conduct the meeting. Do not schedule a meeting that is to last until 4:00 p.m., then permit half the members to leave by 3:00 p.m. to catch the early flight home.

There are other examples. However, let us suggest a framework in which to place these ideas and others. This outline can help identify areas to consider and serve as a basis for gathering additional information. During the next several months, the framework can be refined into a form useful to ATE program managers and project directors.

Characteristics of Effective Advisory Committees

Several principles of effective committees have emerged during the preparation of this chapter. Some of these were found in the literature, and others came from the personal experiences of the authors. A recent study of a national advisory committee, conducted by the evaluator of an ATE center, included some additional ideas.

These are presented here as a first step in improving committee effectiveness.

(1) The purpose and role of the committee should be clearly understood by its members and the group it is supposed to advise.
(2) The structure of the committee is carefully considered in light of its purpose, the project's need, and the resources available. Some of these issues are listed below:
   - Size and composition – The size must be manageable, and it is important to consider the areas of expertise required for the committee.
   - Membership – What is the selection process, and how do you find qualified people?
   - Consultant fees – Most funded project advisory committee members expect to be paid and receive reimbursement for expenses. Projects funded by NSF might face some problems here because of their rather low limit on consultant fees. Other ways of enticing qualified people need to be found, and usually it is the appeal of working on National Science Foundation projects.
   - Meeting times and frequency of meetings – Most advisory committees on funded projects meet only once a year. This is probably too infrequent for

them to be effective. The evaluation advisory committee for the Western Michigan project meets once a year, but several of its members are actively involved in the project throughout the year. They serve on site visit committees, review draft versions of documents, and prepare reports like this one.

Issues have been raised about the potential conflict of interest that might arise when the committee members assume some work of the staff. That is, the possibility exists that they might be biased toward the project when part of the work was theirs. On the other hand, the increased involvement gives committee members a better understanding of what the project is doing. This issue was discussed with NSF staff, and the decision was made to involve committee members in the project's work between meetings. It is still too early to determine if this is a problem because the second meeting of the advisory committee has not yet been held.

- Selecting a chairperson – The importance of this issue was discussed above.
- Staffing and support – Advisory committees are more effective when adequate support is provided to them. It requires time and effort and a person devoted to detail. You can't appoint a committee, have a meeting, and then forget them for a year until the next meeting. Many funded projects lack this capability, and the committees' effectiveness suffers.

Operating policies – A set of procedures to consider for the conduct of meetings needs to be addressed. For example, should the committee follow some prescribed rules of order such as Roberts'? Or are spontaneity and consensus building desirable?

Other issues to consider are techniques for problem solving and decision making, defining the expected output from committee meetings, relationships with project staff, and possible conflicts of interest.

Meeting procedures – Ways to make meetings work.[13]

- Preplanning – Meeting facilities, equipment, and other amenities are suitable and meeting times and places accommodate most of the committee members.
- Effective implementation – This includes promptness, ending on time, staying on task, allowing sufficient time, and permitting full and open discussion. It also should address the problem of the "in and outers," those people who leave periodically to make phone calls, answer e-mail, check their messages, etc.
- Between meetings – It is important to send out the minutes of the meetings, follow up on decisions reached and recommendations made, and distribute the purpose and agenda for the next meeting.

Evaluating the advisory committee – It is important for the project director and staff to conduct periodic evaluations of their advisory committee. If it is found lacking, then the reasons for the problem need to be identified and remedied if possible. Perhaps it is a membership problem, a lack of interest by the committee members, ineffective support for the committee, or a weak leader. If the problem cannot be remedied, the best solution might be to end the committee. No committee at all is probably better than one that is ineffective.

CONCLUDING REMARKS

No doubt other suggestions for effective ATE advisory committee meetings should be considered. These will come from the proposed research study and the expanded literature review mentioned earlier. Then these ideas will be packaged in some way, perhaps a videotape, maybe a Web site, or a handbook. A technique is needed to make the information readily available to principal investigators, NSF, proposal writers, and others involved in organizing and operating effective advisory committees. A videotape showing a "good" and "not so good" committee in operation has its appeal and might be an effective instructional tool.

Whatever the content and whatever the means, the goal of the effort should be to improve the effectiveness of advisory committees, particularly in the ATE program.

POSTSCRIPT

Following the receipt of this report by the National Science Foundation, a supplementary grant was made to The Evaluation Center at Western Michigan University. The purpose of the grant was to further study the advisory committee process as it was being implemented by the ATE projects and centers. Of particular interest were the ATE's National Visiting Committees. These are committees required by NSF for its centers and larger projects.

There were two general tasks for the additional study: (1) to learn more about the advisory committee process in the ATE context; and (2) to develop procedures that can be used to improve the effectiveness of these committees.

Several products have been developed from the additional study. They can be found on The Evaluation Center's Web site at http://www.wmich.edu/evalctr/ate. These documents include two papers authored by Wayne W. Welch and Robert A. Reineke that address the two main questions of the study: *A Study of Advisory Committees Used by ATE Projects and Centers, Final Report-Part A* (August 2002) and *Final Report-Part B* (November 2002).

Part A describes the research procedures and describes the advisory committees used in the ATE program. The second report, Part B, presents suggestions for improving the committee effectiveness. This includes an "Advisory Committee Checklist" for organizing, planning, implementing, and following up a committee. There is also a brochure entitled, "Meeting Reminders for Advisory Committee Members."

A major emphasis of the work was the improvement of National Visiting Committees as used in the ATE program. An NVC Handbook has been developed. This is a checklist with supporting text specifically aimed at the national committees required by NSF. In addition, a Web-based directory of approximately 110 past and current advisors has been produced. This is a resource for new grantees who must organize these committees and is also a means to establish communication among those who are or have served on these committees. Both products can also be found on The Evaluation Center's Web site.

NOTES

1. *Status Report 1: The Nature of the ATE program (2000, May). Status Report 2: Findings from a Survey of ATE Projects and Centers* (2000, October). Kalamazoo: The Evaluation Center, Western Michigan University.
2. The original quote contained the word "children."
3. *Greater Kansas City Medical Bulletin 63*. SUBJECTS: Humankind: Humor and Wit.
4. Other names are advisory panels, steering committees, and national visiting committees. However, their purposes are generally similar.
5. The definition of a "large" project changed in 2002 from one that was funded at $500,000 or more to one funded at $750,000 or more.
6. The brochure's title is "National Visiting Committees for NSF/DUE Sponsored Projects." It is undated but was published in the mid-1990s under a contract to the Montana Systemic Teacher Excellence Program.
7. Teital, L. (1994). *The advisory committee advantage. Creating an effective strategy for programmatic improvement*. Washington, DC: George Washington University.
8. *Ibid.*
9. Fisher, A. (1979). Advisory committees: Does anyone want their advice? *Educational Leadership, 37*(3), 254–255. Axelrod, N. (1991). *Creating and renewing advisory board: Strategies for success*. Washington, DC: National Center for Nonprofit Boards.
10. House Subcommittee on Government Management Information and Technology (1998). *Hearing to examine the use of federal advisory committees*. Washington, DC: Author.
11. The authors wish to thank one of the unnamed reviewers of this chapter for suggesting several of these questions.
12. Smoak, M. G., & Latham, P. L. (1981). *A summary of the impact of advisory committees* (ERIC document ED203170).
13. Some of the ideas in this section come from Mercer, J. W., & Meunier, G. R. (1991). *Fifty indices of effectiveness*. St. Paul, MN: State Council on Vocational Education (ERIC document ED372205).

THE USE OF EVALUATION

Wayne Welch and Frances Lawrenz

ABSTRACT

Evaluation is claimed to be an important component of the Advanced Technology Education (ATE) program. However, an issue exists about the ATE evaluation. Evidence is provided that suggests some evaluation is taking place, however, the information is limited and some findings are contradictory.

Based on the findings, it is recommended that additional information be gathered to fill the knowledge gaps. There is uncertainty about the nature of the evaluation work, who is doing it, and how useful has it been to the projects and centers. A research design is presented to address these and other questions.

INTRODUCTION

The Advanced Technologies Education (ATE) program is a politically visible undertaking by the National Science Foundation (NSF). There was considerable opposition within the Foundation to make grants to community colleges. Claims were made that NSF could do little to affect these colleges because there were so many of them. In addition, there was concern by the research directorates that limited dollars would be drained away from the support of scientific research. However, in 1992, Congress mandated the program and the first grants were made in 1994. By 2002, about 500 grants had been made totaling more than $300 million. The Foundation expects to fund an additional 75 awards totaling $40 million in fiscal 2003.

Evaluating the Upgrading of Technical Courses at Two-Year Colleges:
NSF'S Advanced Technological Education Program
Advances in Program Evaluation, Volume 9, 201–215
Published by Elsevier Ltd.
ISSN: 1474-7863/doi:10.1016/S1474-7863(04)09010-6

Evaluation has been required for ATE grants since its inception although specifics on the nature of the evaluation have been limited. The 1998 Request for Proposals (RFP) for the ATE Program mentioned evaluation only twice. One was, "project evaluation to include, as appropriate, alignment with national standards" (p. 8). The other stated, "Formal proposals for such activities should include evaluation of impact of activity" (p. 9). This statement was included on a list of possible program improvement activities.

The National Science Foundation increased its attention to evaluation at the grant level in the 2000 program solicitation. It included 18 lines describing the kind of evaluation that should be done. However, the 2003 program solicitation only has a passing mention of evaluation. It mentions that an evaluation plan, with other requirements, should be included among the 30 pages permitted for the project description. Evaluation is also mentioned twice under the criteria for proposal review. One question asks reviewers to judge if the evaluation plan is appropriate and adequate and another states "Will the project evaluation inform others through the communication of results?"

Since 1998, all guidelines have included the requirement that all centers and projects[1] with funding levels of $500,000[2] or more should have National Visiting Committees (NVC). These committees have among their charges, to assess the plans and progress of the project and make reports both to the grant leadership and to NSF. This is an additional kind of evaluation procedure required for the ATE program.

Other evaluation activities associated with the ATE program include a grant to the Evaluation Center at Western Michigan University to conduct an evaluation of the national program. This evaluation is designed to assess the impact and effectiveness of the ATE program.

NSF also has produced several documents to assist proposal writers and project directors in conducting evaluations of their programs. There are two "user-friendly" handbooks and an online evaluation resource library.[3] The technology education area is included as a special section in the online library. The early program solicitations mentioned these resources, but the most recent one (2003) does not.

The Foundation's stated interest in evaluation is generally strong. However, little is known about how this interest translates into evaluation practice at the grant level. This issue is examined here using the information gathered by the Western Michigan group. The following questions are addressed. Are grant evaluations being done? If so, are they effective? Are the purposes of the evaluations being achieved, that is, program clarification, improvement and accountability? What changes have occurred because of evaluation findings? Answers to these questions are intended to help decide future ATE evaluation efforts.

PROCEDURES

The Western Michigan evaluation project has conducted several studies and published the results in a series of reports. The first of these, Status Report 1, was published in May 2000. This report was based on a review of program information maintained by NSF, interviews with NSF staff, participation in Principal Investigator (PI) meetings, and "visits" to project web sites. In addition, a sample of 10 grants was selected for a detailed analysis in this report. The sample included two grants from each of the first five years of the program (1994–1998). They were selected to represent a variety of technology fields, type of institution funded, and duration (one to four years). The sample contained four centers and six projects.

Originally, 26 grants were selected for this preliminary analysis. However, the WMU group discovered that examining NSF files, called jackets, is difficult. Some information included in the jackets is confidential and this information is not separated from that which is public. Making the jackets available for inspection requires considerable work by NSF staff. In fact, some divisions and programs refuse to do it because of the time demands. Because of this, there was only enough time available to gather and analyze the materials of 10 of them. Thus, little is known about the information contained in the files for most of the grants.

A second source of information is the reports of a survey of all active grants in 2000 and 2001. Status Report 2 included information from a web-based survey of 113 grants carried out in the spring of 2000. A similar survey was conducted in the spring of 2001. It resulted in a report called, Survey 2001: The Status of ATE Projects and Centers.

The third source of information for this chapter was a set of reports from site visits conducted at 13 projects and centers. A list of about two dozen grants that NSF program officers thought would be willing to participate in a site visit was provided to the WMU team. The team selected 13 from that list to represent the variety of projects and centers currently funded by the ATE program. The selection criteria included size of project, geographic location, technology focus, and type of institution receiving the grant, e.g. two-year college, four-year college.

The site visits were carried out by two or three people contracted by Western Michigan. They included members of the ATE evaluation project, members of its Evaluation Advisory Panel, and external evaluators who have worked with the larger ATE projects and centers for the past several years. The visits were of one or two days duration. A report was prepared and submitted to the grant staff for review. The revised reports were made available to the issue authors.

In 2002, a supplemental grant was made to Western Michigan to study the National Visiting Committees used in the ATE program. Part of that grant was to produce a directory of names that could be used by new ATE grantees to organize their own NVCs. An e-mail survey and two follow-ups were sent to 139 projects and centers asking them for a list of their NVC members. This was the number of active projects at the time who had been in operation for a year or more.

The responses to this survey provide information on NVC use. This has implications for the effectiveness of the NVCs as an additional ATE evaluation activity.

Other sources of information used for this chapter were the ATE program guidelines published by NSF and documents turned over to the authors by the external evaluators of two centers. In addition, NSF program officers provided information about the several evaluative aspects of the program.

Each of the status reports and the 13 site visit reports were searched for references to ATE evaluation. In general, little information on evaluation activity was found in these data sources. Evaluation was not a major focus of these studies. Status Report 1 was primarily a description of the ATE program, a large and complex enterprise. Little information was available on the outcomes of the completed awards and the roles of evaluators were not described.

Status Report 2 was a summary of survey findings and it concentrated on four program elements, collaboration, materials development, professional development and program improvement. Only three questions on the web-based survey directly addressed program evaluation. There were a few questions on student assessment in the "materials development" section of the survey. Three questions asked about pilot testing of materials. One might infer that student assessment was done for program evaluation purposes although this is not clear from the nature of the items.

Describing or analyzing evaluation activity was not a major part of the site visit protocol. The observers were told that the visits should address the nature of the ATE program by describing the various sites. They were to provide information on whether or not the numbers, skills, and knowledge of technicians were changing, and if so, to document any increases in the number of technicians and their skill/knowledge quality. Finally, they were to provide information on how program could be improved and what unintended outcomes may have occurred in the individual projects and centers.

Despite the limited emphasis on evaluation given in the instructions, six of the 13 site visit reports did include a section on the role or use of evaluation. These generally were about two paragraphs long. Three others did not include a separate section on evaluation but devoted approximately two paragraphs to the topic Three site reports referred to evaluation in passing and one did not mention it at all.[4]

The evaluation information found in these sources is described below.

FINDINGS

Status Report 1

The strongest statements on evaluation were found in Status Report 1. Recall that this was a document based on a review of file information maintained by NSF, interviews with NSF staff, participation in Principal Investigator (PI) meetings, and "visits" to project web sites. One major finding of this report was that, "Evaluation is under emphasized in the ATE projects." (p. vi)[5] This conclusion was reached in large part because of the limited information available on the actual outcomes of the projects. Evaluation reports were not included in the information they examined, many projects did not have evaluators, and even when evaluation reports were available they did not include what the reviewers considered, "useful information."

This perception is not shared by the NSF program officers. They have worked with a small group of center evaluators whom they have found very effective. Their evaluation reports are included in the center's annual reports to NSF. Most have worked with the centers for several years. Several of them served on WMU site visit teams and three of them are authors of the issue papers reported in this volume.

Thus, the contradiction. On the one hand, the claim is made that evaluation is under represented but this seems contrary to the experiences that NSF has had with the center evaluators with whom they have worked.[6] What seems needed is more information on the kind of evaluation work that is taking place in the wide range of grants funded by the ATE program.

Survey Status Report 2

In the spring of 2000, a web-based survey was distributed to 113 projects.[7] The last section, labeled, "Part III, Finalization" was answered by 84 (74%) of the respondents and contained three questions about evaluation. They asked whether the project used an evaluator and, if so, when he or she was hired.

About 80% of the respondents reported their site has or had an evaluator. Three-fourths of these were external to the project and approximately one-fourth were internal evaluators. (Similar response patterns were obtained when the survey was administered to a sample of 81 current projects in the spring of 2001.)[8]

The evaluators became involved at various times during the duration of the grants. About one-third helped prepare the initial proposal, another third was hired when the funds became available, 22% joined the project when it was in full operation, and 7% were hired as the project was ending. (Data on this question were not included in the 2001 survey report.)

These items do not provide much information about the nature or extent of the evaluation work. However, they do suggest a much higher involvement in evaluation than was reported in Status Report 1. Only one evaluator found in that 10-project analysis was an external consultant.

The reasons for the apparent discrepancy are unknown. However, in the first instance, the reports contained in NSF files were from older projects, 1999 or earlier. The survey figures were obtained from projects active in the spring of 2000 and 2001. Furthermore, it is one thing to report your project had an evaluator assist with the preparation of the ATE proposal and quite another to find an evaluation report in NSF files. These extremes represent a range of involvement in evaluation for which little information is available.

There is also a potential problem because of the lack of validity information on survey responses. Only about 60% of the grants responded to the evaluation items on the 2000 survey. Also, we don't know if the respondents understood the questions and replied accurately. Many survey items are complex and quite long.

The WMU group worked to improve the survey during the second administration and the response rate was greater. In addition, the response patterns were similar across the two years. This consistency does provide some support for the reliability of the questions that in turn provides some validity evidence.

Another part of the 2000 survey included questions about materials development. A total of 72 respondents indicated they did produce materials.[9] Thirty-nine percent of these reported they assessed student success in comparison to nonparticipating students. While this response does not address evaluation directly, one might infer some of these assessments were part of program evaluation efforts.

These projects also assessed student success on their materials using industry/business standards (49%) and half of them reported they assessed student success using educational standards. (These were not mutually exclusive categories. Respondents were to check all that apply.)

The projects reported they pilot tested (78%) or field tested (89%) their materials but used criteria other than student success. Evidence of success was usually as testimonials, acceptance by other institutions, or reviews by users. However, only four offered what the WMU group thought was concrete evidence of the quality of the materials. The report raises questions about the effectiveness of the pilot and field testing.

These materials development procedures hint at possible evaluation work by the grantees, but questions remain about the use of student assessment data, how the data were gathered, and what methods were used during the pilot test and field-testing stages.

It is important to note that the initial work by the Western Michigan group focused on the major objectives of the ATE program. This included professional development, collaboration, materials development, and program improvement. These issues dictated the selection of the major sections of the survey and the site visit reports. It was only after the data from these studies were obtained that the decision was made to include other issue topics, e.g. sustainability, evaluation, etc. This decision was based on a recommendation by the Evaluation Advisory Panel at its meeting in October 2000. Thus, it is not too surprising that limited information on evaluation was available from the reports of the initial studies.

NVC Directory Survey. As mentioned earlier, a study was made of National Visiting Committees as part of a supplemental grant to the Evaluation Center. Part of the study involved gathering the names of NVC members to include in a directory for new grantees.

Recall that 139 grants were active then. Responses were received from 83 grants, a 60% response rate. (This seems low to us but it is difficult to get PIs to respond unless the inquiry is sent directly from NSF or unless the program officers make a special appeal to answer the questions.) Eventually, we obtained lists of NVC advisors from 29 grants. Eight of these were centers. (One center did not respond to the survey.) The rest of the lists ($N = 21$) were from projects.

Fifty-four of the 83 that replied said they did not have a national advisory committee. About half (54%) of those were from projects funded at $500,000 or less. These grants are not required to have a national committee. However, the other half were considered "larger" projects, yet, they reported they did not have national advisory committees.

We do not know the status of NVC activity among the 56 grants that did not respond to our e-mail inquiry. It doesn't seem much of a task to e-mail a list of the half dozen or so people who serve on your committee. It seems more likely that the failure to respond means they do not have an NVC but we are not sure. Nevertheless, the scope of evaluation activity as represented by the National Visiting Committees is limited. There are several very active NVCs, especially at the centers and among a few of the larger projects. However, they are scarce among the projects, even those with funding levels above $500,000.

There is some evidence that most projects have local or regional advisory committees but little is known about their evaluation efforts. Such committees are a common part of community colleges.[10]

Site Visit Findings

Each of the 13 site visit reports was examined for information about evaluation activity at that site. Twenty-seven paragraphs were found in the 12 reports that mentioned evaluation. The 13th report did not mention evaluation, however, a copy of an evaluation report for that project was made available to the authors.

A search was made of the reports using the key word "evaluation." The comments were listed, counted, and grouped into three categories; Uses of evaluation, Evaluation findings, and Recommendations made by the site visit team. The results are listed below.

USES OF EVALUATION

The ways in which evaluation was used in the 13 projects and centers are listed by decreasing frequency of mention. These are the numbers in the parentheses.

(1) Developed and implemented surveys of workshops and institute satisfaction (8).
(2) Gathered data to meet NSF reporting requirements (4).
(3) External evaluators described program and documented program progress (4).
(4) National visiting committees reviewed progress and made recommendations (4).
(5) Informal evaluation of materials development by professional judgment (3).
(6) Developed and presented an evaluation plan (3).

A few other uses were mentioned in the site reports but none was cited more than once and are not reported here.

Participant questionnaires were the most common type of evaluation technique reported by the site visit teams. Documenting progress and reporting data were also mentioned in the site visit reports. Only three teams found evidence that a formal evaluation plan was followed.

Four of the 13 (31%) reported they used external evaluators. This is less than the 75% having external evaluators as reported on the surveys. This comparison is a little hazardous. Thirteen sites is a small sample giving the 31% figure a large standard error.

Also, the survey had the usual non response problems and the 75% figure is valid only if one assumes the non responding numbers were similar to those who responded. The project did a small non response bias study and found no evidence of bias. However, one makes the generalization with caution. Although there may be statistical explanations for the difference in the two figures, there is a substantial contradiction in the figures obtain from the site visits and from the survey.

EVALUATION FINDINGS

Little information was provided in the site visit reports on the findings from evaluation studies. Three sites mentioned that the projects were "on track," a type of evaluative judgment, but without any documentation.

Two reports stated that the participants attending workshops were satisfied with the experiences. One concluded that a course was a success because the students passed a subsequent certification examination.

Because of the focus of the site visits, one might not expect much in the way of evaluation findings. On the other hand, it may be that there are few projects who conduct formal evaluation studies of their grant. We know in one site visit that the first author attended, an evaluation report did not have much of an impact on the project. The project had in its files two evaluation reports, one from an external evaluator and one from a national visiting committee. However, these were not even mentioned until the last half hour of our visit and it was clear the staff was not very familiar with them.

EVALUATION RECOMMENDATIONS

According to the site visitors, several limitations were noted in the evaluations they did analyze. These limitations lead to several recommendations. These are listed by the frequency of mention.

(1) Need more outcome data, especially student impact data (7).
(2) The standards for materials development should be used (4).
(3) A large scale evaluation plan is needed (2).
(4) Modules should be evaluated during the development phase (2).

A few of the reports mentioned that the projects intended to do more evaluation during the coming year. For example, one project reported to the site visit team that it intended to "promote and improve evaluation efforts." However, no information is available to determine if this occurred.

It is well to remind the reader that the above comments were based on a purposeful sample of 13 sites from a population of 113 that were in operation in the spring of 2000. It says nothing about the other 137 projects that had completed their work nor projects and centers funded in fiscal 2001.

External Evaluator Activities

Three of 11 ATE center evaluators participated in the site visits. They were asked about the nature of the work they did. The evaluation plans and an annual report for two of the Centers were made available to the authors. Both evaluators have worked with their center for more than five years. The evaluation reports are submitted to the Center staff and attached to the annual reports provided to NSF. Just how NSF uses these reports is not clear.

One evaluation plan reviewed was particularly well designed. The purposes of the evaluation were clearly stated and appropriate evaluation questions were posed. A variety of methods were used to gather the information to answer these questions. Findings of the previous year's evaluation were presented and plans for the coming year were described.

The other evaluation plan we examined addressed some evaluation issues but mostly it was designed to provide the data the center needs to complete its reporting requirements to NSF. The information is provided to the evaluator by the project staff and he summarizes it for the center's report.

The plan also describes some evaluation activities that will be conducted besides gathering the data needed for reporting requirements. Student satisfaction with their courses and employer satisfaction with graduates of the program were to be assessed. They planned to use surveys, interviews, and focus groups to gather the information. Specific evaluation questions were not posed.

Some Impressions

Several impressions on the role of evaluation in the ATE program emerged during this review. These opinions are based on the information gathered by the Western Michigan group and from authors' experiences as a Co-PI, a consultant to the project, the chair of its evaluation advisory panel, and as members of the site visit teams.

(1) Most projects and centers pay some attention to evaluation but little is known about the precise nature of the work being done. There is contradictory information on the evaluation work.
(2) Evaluation is probably included in the grants because it is mandated by NSF but one does not sense the projects feel a strong need for it.
(3) When evaluation is carried out, it consists of documenting project activities, post treatment questionnaires, and the gathering of data for NSF yearly reports.
(4) No evidence of measuring success using student scores was found although many site visit teams thought this should be done.
(5) Much material development work is carried out without any formal evaluation process. Pilot testing and field testing were limited and heavy reliance was placed on professional opinion.
(6) Little is known about the usefulness of advisory boards and national visiting committees and how their work fits with that done by project evaluators, both internal and external.

(7) Not much information is available about evaluation findings and their impact upon the ATE grants and the national program.
(8) Investigating the costs and benefits of conducting program evaluations seems prudent. Should evaluation be expanded? Reduced? Perhaps eliminated?

Given the gaps in our understanding of evaluation activities in the ATE program, additional investigation in this area seems warranted. A line of research is proposed in the next section.

RECOMMENDATIONS

Evaluation and the ATE Program

The evaluation and oversight of the national program includes the Government Performance and Results Act (GRPA) of 1993, three-year reviews by a Committee of Visitors, and the Western Michigan University study. There are also mandated evaluations at the grant level. However, the evidence that such evaluations are effective is contradictory. The project directors report they use evaluators but an investigation of the NSF files revealed few evaluation reports. The site visit reports included evidence that some evaluation work is occurring but questions were raised about its effectiveness. The survey responses indicate that evaluators are used at the projects and centers. However, we know little about what evaluation is done, who does it, and how it is used. We do not know if the evaluation efforts are perceived as useful by the grants or future grantees, nor do we have any information on cost effectiveness.

Our current knowledge base is inadequate to address these issues. Neither the information gathered thus far by the Western Michigan group nor the information available from NSF is sufficient to address the purposes of this chapter. Namely, to make decisions or reach conclusions about the use of evaluation in the ATE program.

Because of the need for additional information, several research questions proposed for the Western Michigan evaluation study.[11] These are listed below.

Evaluation Research Questions

(1) Which projects and centers carry out evaluations?
(2) Who does the evaluations?
(3) What kinds of studies are done?
(4) What information is obtained from the studies?
(5) How is the evaluation information used?

(6) Does it improve the quality of the programs?
(7) What is the cost of doing evaluations?
(8) Are the studies perceived as useful?
(9) How have evaluations affected the grants and ATE program?
(10) What are the strengths and limitations of the evaluation work?

To address the research questions, certain kinds of data need to be gathered. Table 1 includes the questions, information needed to answer the questions, and methods by which this information could be gathered.

It can be seen from an examination of Table 1 that information must be obtained from PIs, their staff, evaluators (if they had them), NSF, and possibly site visits. The question arises, "How best gather that information?" The answer to this question is the basis for the research design of the study.

After considering several alternatives, and aware of the response burden already on the grantees, the following research design is offered.

Table 1.

Evaluation Research Question	Information Needed	Source of Information
1. Which projects and centers carry out evaluations?	A list of the grants that did evaluation plus their demographic characteristics	Project directors or NSF files
2. Who does the evaluations?	A list of evaluators plus their demographics	Project directors plus confirmation from the evaluators
3. What kinds of studies are done?	Copies of evaluation plans and reports	Project directors and evaluators
4. What information is obtained from the studies?	Evaluation reports, verbal descriptions of actual studies	Project directors and evaluators
5. How is the evaluation information used?	Perceptions or documentation of use	Project directors, staff, evaluators, advisory committees, NSF staff
6. Does it improve the quality of the programs?	Documented changes, perceptions	Project reports, project directors, staff, evaluators
7. What is the cost of doing evaluations?	Budgeted and actual expenses for evaluation work	Project directors, NSF expense reports
8. Are the studies perceived as useful?	Opinions about the evaluations	Project staff, evaluators, NSF staff
9. How have evaluations affected the grants and the ATE program?	Opinions and documented evidence	Project staff, evaluators, NSF, perhaps site visit committees
10. What are the strengths and limitations of the evaluation work?	Judgments based on information gathered in questions 1–9	Issue chapter authors and reviewers

RESEARCH DESIGN

The population of projects and centers funded during a three-year period is about 125, 115 projects, ten centers. A random sample of 30, 10 from each year, should provide sufficient data to answer the questions posed above. A semi-structured telephone interview may be an effective means to obtain the necessary information. A response rate of 80% or more would be the goal so that one can generalize to the total population with a high degree of certainty.

Follow up telephone surveys would be made to the evaluators for the grants that had them. The questions would be written to obtain the information described in Table 1. This procedure works well when the questions are sent to the respondents before the phone call.

The steps in the research process would be:

(1) Identify the projects and centers funded between fiscal 2000 and 2002.
(2) Draw a stratified random sample. Strata would include projects/centers and funding years. Eight projects and two centers would be drawn from each of the three years.
(3) Obtain names, addresses, and phone numbers of principal investigators of the sampled sites.
(4) Draft semi-structured interview protocols for PIs, evaluators, and any others needed to answer the research questions.
(5) Pilot test the interview on a sample of three sites not in the random sample. Revise as appropriate.
(6) Draft a letter of explanation and invitation to the project directors and evaluators. This letter should include a request for information on the best time to contact them by telephone.
(7) Pilot test the letter on three non sampled sites and make necessary changes.
(8) Implement the study.
(9) Conduct follow up telephone calls as necessary to obtain an 80% response rate.
(10) Analyze data and prepare a report.

Some suggestions for the kind of questions to be asked during the telephone interview are:

(a) Did your project have an evaluator? If so, what was the name of the evaluator or the evaluation team?
(b) What kind of information was gathered?
(c) When was the evaluation carried out?
(d) What was the duration of the evaluation?

(e) How much money did the evaluation cost? Include salary or fees and expenses.
(f) Was the evaluation completed on time?
(g) Was the information useful? If so, in what ways?

Also, a few direct questions should be asked to probe their attitudes toward evaluation. For example, we might ask respondents to Agree or Disagree with the following statements.

Evaluation activities are well worth the money we spent on them.
There is too much time and effort being devoted to project evaluations.
We know what to do. We do not need some outsider telling us what we should be doing.
Evaluations are a waste of time.
One of the best things about the ATE program is its attention to evaluation.

The semi-structured interview is the key part of this research study. The proposed domain sketched out in Table 1 should be carefully reviewed and revised by the WMU staff and their consultants. A pool of items should be written and pilot tested. The interviews should be limited to 30 minutes, so each question must be carefully crafted to obtain the information required.

We believe the study described above is doable without a heavy response burden on the staffs of the ATE grants. It should provide an important contribution to our understanding of the role of evaluation and the ATE program. It will add to our understanding of evaluation in general, and the findings and recommendations should have implications for the range of programs funded by the Education Directorate of the National Science Foundation.

POSTSCRIPT

As of the date of the publication of this volume, the above study has not been funded nor has it been proposed to NSF. However, several questions listed above have been recommended for inclusion in the next grantee survey, Survey 2004.

NOTES

1. Centers are comprehensive in scope with funding levels of \$1.5–\$4 million. Projects usually focus on a specific activity, for example, professional development. Funding for three years is usually less than \$1 million.
2. This figure was increased to \$750,000 in 2002.
3. The online library is available at oerl@www.ctl.sri.com.

4. Although the site report did not include a discussion of evaluation, the site visitors were provided with a report prepared by their external evaluator. This was made available to the authors.

5. This section of the Status Report 1 uses the non italicized word "projects" to include both centers and projects. Based on the size of the grant, it appears the sample they were discussing included four centers and six projects.

6. Recall that there are only 11 centers among the approximately 125 *grants* currently in operation.

7. See Status Report 2 for an explanation of the survey sample and process.

8. Survey 2001. (July 2001). The Status of ATE Projects and Centers. Kalmazoo, MI: The Evaluation Center. Western Michigan University. (Draft).

9. Some projects were not involved in materials development and others did not respond to this part of the survey.

10. See the chapter in this volume by Wayne Welch and Arlen Gullickson entitled "The Use of Advisory Committees in the ATE Program."

11. The questions are not meant to be exhaustive. It seems likely that others will emerge as the research plan is refined and implemented.

ABOUT THE AUTHORS

Norman Gold has funded and conducted major studies in education reform and improvement over the past 30 years. Over the past decade, he has focused his studies on developing technology capacity in education, with professional development being one of the key components. He is a member of the Western Michigan University's (WMU) evaluation project's evaluator group and conducted site visits for this project.

Arlen R. Gullickson is Principal Investigator/Project Director for the ATE evaluation project at The Evaluation Center at Western Michigan University. He is the Director of The Evaluation Center and serves as project director for other NSF projects.

Nanette Keiser is the Co-PI/Project Manager of the ATE evaluation project. She was involved in the design, implementation, data collection, and data analysis phases of the survey and site visits. She served as a Senior Research Associate at WMU's Evaluation Center.

Frances Lawrenz is the Co-Principal Investigator and Senior Associate of the ATE Evaluation Project. She was involved in the design, implementation, data collection, and data analysis phases of the survey and site visits. She is the Chair of the Department of Educational Psychology at the University of Minnesota.

Thomas Owens is Associate Director of the Education and Work Program at the NorthWest Regional Educational Laboratory (NWREL). He has served as an external evaluator of the National Workforce Center for Emerging Technologies (NWCET) for the past five years. He has also been a consultant to the WMU evaluation project and a member of its evaluator group.

Karen Powe is a writer and education consultant with award-winning publications and 18 years' experience in education policy and program development. As a trainer for education policymakers and administrators, she has worked at local, national, and international levels.

Lester W. Reed, Jr., Professor of Education and Senior Associate at the Western Center for Community College Development at Oregon State University, has more than 25 years' experience in the area of technical education. During the last 6 years, he has served as the external evaluator for an ATE Center of Excellence and various projects. He has also been a member of the evaluator group for the WMU evaluation project and conducted site visits for this project.

Gloria Rogers is vice President for Institutional Research, Planning and Assessment at Rose-Hulman Institute of Technology. She has been an external evaluator for the ATE center, the "National Center of Excellence for Advanced Manufacturing Education" at Sinclair Community College, and a project entitled "Building a National Employer-Based Technical Education System" administered by Educational Development Center, Inc. She also served as a member of the evaluator group for the WMU evaluation project and conducted site visits for this project.

Gloria R. Tressler was a Research Assistant for the ATE evaluation project at the WMU Evaluation Center.

Wayne W. Welch is a Professor Emeritus of Educational Psychology at the University of Minnesota. His major areas of interest are program evaluation and science education. He has conducted more than 60 evaluation studies for foundations, schools, and health and human service agencies such as the National Science Foundation, American Cancer Society, and the St. Paul Foundation. He also worked at the National Science Foundation in 1972–1973 and 1988–1989 when he headed their Office of Studies and Program Assessment. In1995 he received the Distinguished Research Contribution Award of the National Association for Research in Science Teaching.